PLANT THEORY

PLANT THEORY

BIOPOWER & VEGETABLE LIFE

JEFFREY T. NEALON

STANFORD UNIVERSITY PRESS   STANFORD, CALIFORNIA

Stanford University Press
Stanford, California

Printed in the United States of America on acid-free, archival-quality paper

Library of Congress Cataloging-in-Publication Data

Nealon, Jeffrey T. (Jeffrey Thomas), author.
  Plant theory : biopower and vegetable life / Jeffrey T. Nealon.
    pages cm
  Includes bibliographical references and index.
  This book joins the growing philosophical literature on vegetable life to ask what changes in our present humanities debates about biopower and Animal Studies if we take plants as a linchpin for thinking about biopolitics.
    ISBN 978-0-8047-9571-5 (cloth : alk. paper) — ISBN 978-0-8047-9675-0 (pbk. : alk. paper)
  1. Plants (Philosophy) 2. Biopolitics. I. Title.
  B105.P535N43 2015
  113—dc23
                              2015021846
ISBN 9-780-8047-9678-1 (electronic)

Typeset by Bruce Lundquist in 10/14 Minion

For Rich Doyle
Il miglior fabbro

CONTENTS

Preface: Plant Theory?   ix

Acknowledgments   xvii

Abbreviations   xix

1. The First Birth of Biopower:
    From Plant to Animal Life in Foucault   1

2. Thinking Plants with Aristotle and Heidegger   29

3. Animal and Plant, Life and World in Derrida;
    or, The Plant and the Sovereign   49

4. From the World to the Territory:
    Vegetable Life in Deleuze and Guattari;
    or, What Is a Rhizome?   83

    Coda: What Difference Does It Make?   109

    Notes   123

    Index   141

PREFACE

Plant Theory?

The genesis of this obscure-sounding project was actually quite simple: it started a few years ago at the annual Modern Language Association conference, where it seemed to me that an inordinate number of panels were dedicated to discussions of animal life in literature, culture, and theory. In 2009 a special cluster in the MLA's flagship journal, *PMLA*, heralded the official arrival of animal studies in the literary humanities, and a torrent of books and articles has ensued. Going forward, it looks as if strong interest in animals among humanities scholars will continue, as there are a number of established book series and journals now dedicated to the topic. And while I like animals as much as the next person, the Foucauldian in me became preoccupied with trying to figure out how animality had somehow become the "next big thing" in the world of humanities theory and criticism. I wanted to figure out how, to paraphrase Michel Foucault, a "today" saturated with animal studies was different from a "yesterday" when, at least in the circles that I frequented in the humanities, not too many people were thinking and writing about animals. Indeed, if you had wagered in the mid-1990s that within a decade animal life was going to be an incredibly hot topic in literary and cultural theory (that there would be MLA sessions dedicated to animals in the Middle Ages, in Faulkner, or even a panel on "Cetacean Nations," which included a presentation on narwhals), most humanities academics would gladly have accepted that bet, and offered long odds as well.[1]

Nevertheless, on reflection there remains plenty of continuity within this story of rapidly shifting paradigms, insofar as the theoretical discourse surrounding animal lives emerged on the scaffolding of the Big Theory era's master thinkers. Most obviously, Foucault's work has been key in coming to grips with biopower (how various practices and concepts of "life" became central for power and knowledge in the modern and postmodern era), and Jacques Derrida remains a central theorist of animal studies, having dedicated several long texts to the question of animality in the canon of Western thinking. Likewise,

Gilles Deleuze and Felix Guattari thoroughly rethought the question of life through the traditions of vitalism, and along the way they offered animal studies one of its founding concepts: "becoming-animal." (The "Cetacean Nations" MLA panel, for example, included a paper on "becoming-whale.")

So on the heels of my earlier work on Foucault, I first set out to do a kind of genealogy of animal studies in the humanities, tracing its emergence out of the Big Theory era of the 1980s and 1990s. I specifically began by looking to untangle animal studies' relations to the triumph of biopower (which filters everything through the concerns and practices of human life), starting from the plausible hypothesis that animal studies' ascendency would turn out to be a natural outgrowth of the Foucauldian critique of biopower, a way for the humanities to begin undermining its myopic focus on human life as the only form of life worth the name. I expected to find out that animal studies became central in our era of intensified biopower precisely because animal life constitutes a visible and privileged "other" to human life: animals become a locus for critique because they lead lives that *don't* count within a biopolitical regime that's otherwise obsessively focused on questions of life. In short, my initial hypothesis was that Foucault might show us how animals' exclusion from an increasingly triumphant human biopower was precisely what had made animals such ethically compelling figures over the past decade.

As Chapter 1 outlines, however, I did not find a confirmation of that hypothesis when turning to Foucault to unravel these problems of life and animality. Foucault in fact provides a very different explanation for intensified interest in animals within the biopolitical era: biopolitics remains invested in animals not because animals constitute our "others" but because animality provides the subtending notion of subjective desire that gives rise to biopower in the first place. Foucault argues that animals—and their hidden life of desire—have from the beginning been the privileged figures for understanding human life within the regime of biopower. Even more intriguingly, Foucault argues that it's not animality but a primary focus on plant life that gets left behind in the era of biopower. In other words it became clear to me that the plant, rather than the animal, functions as that form of life forgotten and abjected within a dominant regime of humanist biopower.

I then turned to animal studies, and to Derrida in particular, to make sense of that abjection or exclusion of vegetable life within the voluminous work on nonhuman forms of life; instead, I found there an even more stubborn series of elisions. I found a pattern of dedicated swerves around the question of vegetable life in Derrida's work on animals, as well as the Heideggerian work on "world" and "life" that remains so central to Derrida's thinking on these topics. As we will see, several times both Derrida and Heidegger (not to mention Agamben) bump up against the question of vegetable life within their extensive work on humans and animals, but just as quickly and decisively, they each continue to elide the question of vegetable life, to remain focused on humans and animals. I began seeing a familiar gesture—consistently sidestepping vegetable life within the theoretical discussion of life-forms—and this led the Derridean in me to believe that I was on to something that I needed to follow out. And following this trail of vegetable life led me finally to Deleuze and Guattari, as the most famous practitioners of a plant-based rhizomatic mode of thought.

Even more surprising than that elision of plant life in recent biopolitical theory, however, was the amount of outright hostility toward thinking about vegetable life that I found within animal studies itself. For example, Cary Wolfe's *Before the Law: Humans and Animals in a Biopolitical Frame* ends with a very dismissive sense of skepticism about plants' claims to the hard-fought ethical gains of animal studies. He calls such sentiments a "cop out" and a "refusal to take seriously the differences between different forms of life—sunflowers versus bonobos."[2] Clearly, plants have no place in the "biopolitical frame" that remains reserved, it seems, for "humans and animals."

Likewise, Gary Francione, the coeditor of Columbia University Press's "Critical Perspectives on Animals" series, is having none of this vegetable nonsense. In a debate with Michael Marder, author of the incisive *Plant-Thinking: A Philosophy of Vegetable Life*, Francione insists that "There is . . . not one shred of evidence about which I am aware that plants suffer or have any intentional states," so they have no "interests" and are not entitled to ethical recognition or any form of "subjectivity." Perhaps more contentiously, Francione adds:

> I should note in the 30 years I have been doing this work [in animal studies], when I discuss this issue with people who are not vegans, the conversation almost invariably turns to a sudden solicitude for the "interests" of the vegetables on our plates. We both know that the primary audience for your book will not be vegans who want to ponder whether they are under-inclusive ethically, but those who claim that we should skip over the interests of the cow and worry about whether the carrot had a tough harvesting season. If . . . this enterprise is really about putting cows and corn in the same group, then it would most certainly be an attempt to undermine veganism.[3]

And I would add that both Francione and Wolfe are incredibly kind compared
to the vitriolic fare to be found in the comment lines under numerous recent
newspaper stories and blog posts concerning the growing scientific consensus
around "plant intelligence." On a more personal register, I've had a vegetarian
academic friend (ex-friend?) berate the idea that plants are compellingly alive
as, I'm quoting here, "something a stupid frat guy would say." As perhaps the
ultimate backhanded dismissal, as of this writing (January 2015), the Wikipedia
page on "Plants Rights" quotes Oklahoma City bomber Timothy McVeigh as
a strong defender of plant lives (suggesting that not only frat guys but mass
murderers endorse this ridiculous idea that plants are worth thinking about—
though to be fair, Wikipedia also has an extensive entry on Adolph Hitler's strict
vegetarianism). Of course, *New York Times* articles with vegetarian-baiting ti-
tles like "Sorry, Vegans, Brussels Sprouts Like to Live, Too" and "No Face, but
Plants Like Life Too,"[4] might help you understand why some animal studies
practitioners get upset when the discussion turns to plant life; but from another
angle, and debates about ethical vegetarianism aside, animal studies' blanket refusal to consider vegetable life within its biopolitical frame seems to function as
a subset of an old practice: trying to close the barn door of ethical consideration
right after your chosen group has gotten out of the cold of historical neglect.

What do I mean by that? In *Gender Trouble* (1990) Judith Butler pointed
out that the era of identity politics was haunted by an "embarrassed etc." that
inevitably attaches itself to any list of social recognition categories: race, class,
gender, sexual orientation, disability, age, etc. Wendy Brown's *States of Injury*
(1995) offered the additional insight that subaltern status or "injury" had become the coin of the realm for recognition claims within identity politics. In
the wake of those powerful analyses, it was really only a matter of time before
both the "etc." and the claims to "injury" decisively jumped the species line:
why restrict the recognition of subaltern status to human lives? Cue, then, a
robust emergent literature in animal studies.

If you have been paying attention to that recent flood of work in animal
studies (and I guess even if you haven't), a certain kind of "critical plant studies" might strike you as the logical next step in the conversation about ethically
compelling forms and modalities of life. Indeed, while animal life has been
othered, ignored, or disrespected in most Western thinking, vegetal life has
had an even rougher go of it. Saint Thomas Aquinas, that *bête noir* of animal
studies, concisely sums up the philosophical prejudice: "Even brute animals are
more noble than plants."[5]

But I'll stress right out of the chute that in *Plant Theory* I'm not primarily trying to put plants on the humanities' ethical, political, or theoretical map, at least in part because they're already all over that terrain so decisively. Despite skepticism concerning the questions posed by vegetable life, there exists the beginnings of a theoretical literature sprouting up around plant life and biopolitics: in addition to Marder's sweeping philosophical work in *Plant-Thinking*, there's Richard Doyle's tour de force *Darwin's Pharmacy: Plants, Sex, and the Noosphere*, Matthew Hall's political theory turn in *Plants as Persons*, Elaine Miller's feminist account of German romanticism in *Vegetative Soul: From Philosophy of Nature to Subjectivity in the Feminine*, or Eduardo Kohn's posthuman anthropology *How Forests Think*, not to mention a vast environmentalist literature that takes the question of life beyond the human-animal divide very seriously indeed—everything from Timothy Morton's groundbreaking *Ecology Without Nature*, through theoretical work on climate change (like Claire Colebrook's sharp *Essays on Extinction*), via a detour through early modern studies (many of the essays in Jeffrey Jerome Cohen's collection *Animal, Vegetable, Mineral: Ethics and Objects*) and studies in romanticism (for example, Theresa Kelley's *Clandestine Marriage: Botany and Romantic Culture* and Robert Mitchell's *Experimental Life: Vitalism in Romantic Science and Literature*).[6] All of these books are scholarly accounts published by university presses, but the winding road of plant theory also leads back to more practical, movement- and rights-based treatments of the earth, the environment, and the battle over genetically modified plants or the politics of food production (important exposés like Marie-Monique Robin's *The World According to Monsanto*, Wenonah Hauter's *Foodopoly: The Battle over the Future of Food*, and Frederick Kaufmann's *Bet the Farm: How Food Stopped Being Food*).

Likewise, over the past few years there has emerged an especially vigorous popular critical literature dedicated to translating the emergent hard-sciences consensus concerning "plant intelligence," or what botanist Anthony Trewavas straightforwardly calls "plant behaviour."[7] In addition to Daniel Chamovitz's *What a Plant Knows*, Michael Pollan's *The Botany of Desire: A Plant's-Eye View of the World*, and Francis Halle's *In Praise of Plants*, I've already noted several recent articles on plant intelligence in the *New York Times*, as well as extended feature pieces by Oliver Sacks in the *New York Review of Books* and by Pollan in *The New Yorker*.[8] Chamovitz sums up this recent research confirming that "plants do indeed have senses":

> Plants are acutely aware of the world around them. They are aware of their visual environment; they differentiate between red, blue, far-red, and UV lights and

respond accordingly. They are aware of aromas surrounding them and respond
to minute quantities of volatile compounds wafting in the air. Plants know when
they are being touched and can distinguish different touches. They are aware
of gravity: they can change their shapes to ensure that roots grow up and roots
grow down. And plants are aware of their past: they remember past infections
and the conditions they've weathered and then modify their current physiology
based on those memories. . . . What we must see is that on a broad level we share
biology not only with chimps and dogs but also with begonias and sequoias.[9]

As a kind of clincher for the surge in interest surrounding plant thought
and action, even a summer 2014 episode of *America's Got Talent* had a brush
with vegetable greatness, featuring a woman who hooked up some plants to a
lie-detector-style apparatus to track their electromagnetic energy and make
beautiful music. (Several plant-generated music CDs are already available
from DataGarden.com, and the relatively simple PsychoGalvanometer necessary to produce the music is now available for sale, as the MIDI Sprout).[10] Even
your Xbox is onto plant agency, which is prominently featured in the popular
all-ages first-person shooter game *Plants vs. Zombies*. (These may seem like
odd adversaries for a combat game, but the appeal to the whole family is clear:
both plants and zombies are swarms of quasi life that one can kill with a clear
conscience, insofar as they're only liminally alive in the first place.) Finally, a
plant in Japan, Midori-san, had its own blog, with algorithmic software designed to translate its photosynthesis and respiration into affective statements
like this: "Today was a sunny day and I was able to sunbathe a lot."[11] Not exactly Nabokov, I admit, but passable Internet fare, even though millions of humans presently updating their Facebook pages with similarly banal sentiments
would undoubtedly object to being characterized as house plants. Debates
about plants as musicians or bloggers or zombie killers notwithstanding, I'm
merely trying to highlight the fact that, both inside and outside the academy,
the questions posed by "life" outside or beyond the narrow confines of the
human remain central in today's posthuman, ecocentric, climate-threatened,
locavore world.

Plants, one might say in a kind of cryptic shorthand, are quickly becoming
the new animals. And more power to them. However, my interests here will lie
less in lauding plants or wondering about animals and will remain more directed at a theoretical inquiry into vegetable life's liminal place within the wider
biopolitical focus on "life" in humanities theory today. My primary task is not
to convince you, Dear Reader, to be interested in the biopolitics of vegetable life

(because as I've noted, that interest already exists). I will rather concern myself with the question of what (if anything) changes in our present humanities debates about animal studies and biopower if we take vegetable life into account or if we take plants to be a linchpin for thinking about biopolitics? Even more specifically, I will focus on the roles of vegetable life in the extant high theory of a generation ago (especially Foucault, Derrida, and Deleuze and Guattari), in an attempt to suggest genealogical continuities, breaks, and roads not taken within the recent theoretical past.

It may seem banal to point it out, but within all the discussion of biopolitics and posthumanism, there seems a lot more discussion on the "politics" end of biopolitics than there is rumination on the "bio" part of the story.[12] Perhaps the political questions—animal rights, climate change, mass extinctions—are simply too pressing to allow the luxury of much time to think through and critique the constituent parts of *biopower*. But it's just that kind of slowing down that interests me here and why I return to the high theory of Foucault, Derrida, and Deleuze and Guattari in this urgent era of ecological disaster: because I think we need to rethink these questions, about life, from the ground up.

This project is, finally, aimed not at recuperating that golden theory past but at imagining possible futures for an ongoing and robust biopolitical debate. If, as Eugene Thacker insists, "What we need is a *critique* of life"[13] rather than a celebration or denunciation of our intensified era of biopower, then the critical tools we've inherited from Foucault, Derrida, and Deleuze and Guattari should be crucial in helping us to further that critique. I'll likewise be arguing throughout that as a mode of thought and inquiry in the humanities, "theory" remains alive and well. And with its turns toward biopower and animal studies, humanities theory has trained its sights on the crucial (and "precarious," as Judith Butler insists)[14] question of "life" in the present and going forward. In the end *Plant Theory* suggests that the discourses of contemporary biopolitics may just need a little water and sunlight, and we likewise need to do some turning of the theoretical soil in which the biopolitics debate originally grew—Foucault, Derrida, and Deleuze and Guattari. Going forward, the biopolitics debate will need to take into account an even more robust notion of what constitutes "life" beyond the human.

ACKNOWLEDGMENTS

If this book constitutes a kind of plateau in the manner of Deleuze and Guattari's *A Thousand Plateaus*, its date of emergence would have to be 6 January 2012, in a very long bourbon-fueled discussion at a bar in Seattle with Marco Abel, Gregg Lambert, and Cary Wolfe. Thanks to them for their help in marking out this territory and enduring this book's arguments in a series of disorganized forms. Claire Colebrook, Grant Farred, Lynne Huffer, David Krell, Michael Naas, and Dennis Schmidt also offered crucial insight and encouragement along the way.

This book was largely written from the ground up on my sabbatical year, so special thanks are due to Susan Welch, Dean of the College of Liberal Arts at Penn State, and Michael Bérubé, Director of the Penn State Institute for the Arts and Humanities: the gift of time should never be taken lightly. Thanks also to all the participants in the graduate seminar I taught at Penn State on this topic; they helped me refine and rethink everything.

I also got a ton of very helpful feedback when delivering parts of this book as talks at Louisiana State University (thanks to John Protevi), the University of Louisville (thanks especially to Stephen Schneider and Aaron Jaffe), and the Netherlands Institute for Cultural Analysis at the University of Amsterdam (where I got incredibly valuable feedback, as well as plenty of Dutch hospitality, from Iris van der Tuin and Gaston Franssen). Several meetings of the Secret Society for Biopolitical Futures have likewise sharpened this book immensely: as for the names of specific participants, you know who you are.

My editor at Stanford University Press, Emily-Jane Cohen, was fantastic as usual, offering valuable feedback and invaluable encouragement. And Joe Abbott did a very fine job copyediting the manuscript.

And of course Leisha, Bram, and Dash remain the reasons why any of this matters.

In the end, though, this book would have been impossible for me to conceive or write without the everyday provocation, friendship, and insight of Rich Doyle, who's convinced me over the past decade of plant life's crucial biopolitical importance—among so many other things. His forceful understanding of plant life is all over every page herein. At the end of the day Rich is certainly "the better craftsman" of plant theory, and this book is for him.

ABBREVIATIONS

AK   *The Archaeology of Knowledge*, by Michel Foucault

ATT   *The Animal That Therefore I Am*, by Jacques Derrida

BS   *The Beast and the Sovereign*, 2 vols., by Jacques Derrida

BT   *Being and Time*, by Martin Heidegger

DA   *De anima*, by Aristotle

DR   *Difference and Repetition*, by Gilles Deleuze

G   *Glas*, by Jacques Derrida

HM   *History of Madness*, by Michel Foucault

HS   *Homo Sacer*, by Giorgio Agamben

OT   *The Order of Things*, by Michel Foucault

PA   *Parts of Animals*, by Aristotle

SZ   *Sein und Zeit*, by Martin Heidegger

TP   *A Thousand Plateaus*, vol. 2, by Félix Guattari and Gilles Deleuze

WFS   *The Fundamental Concepts of Metaphysics*, by Martin Heidegger

"WM"   "White Mythology," by Jacques Derrida

WP   *What Is Philosophy?* by Gilles Deleuze and Félix Guattari

PLANT THEORY

1

THE FIRST BIRTH OF BIOPOWER

From Plant to Animal Life in Foucault

IF CROSS-DISCIPLINARY MOVEMENTS in the North American university function like financial instruments (which, of course, they do and they don't), the strongest "buy" orders of recent years would have to come from the burgeoning discourse surrounding biopower and the related body of work dedicated to animal studies. I suppose this book itself attests to continuing scholarly interest in biopower. And animal studies, for its part, has made a strong pitch to be the "next big thing" in the academy, or so the *New York Times* has announced.[1]

These two emerging fields of practice are, of course, intimately related: if biopolitical studies began by pointing out that questions pertaining to human "life" have become *the* political topics of the modern era (revolving around practices of identity, health, and sexuality), animal studies steps in to show how that notion of human-centered biopower is itself based on an originary exclusion and abjection of its other, animal life. In classic deconstructive form (Jacques Derrida is in fact one of the most often cited figures), animal studies shows how the privileged term of biopower (human life itself) is made possible and remains hegemonic through its illegitimate forgetting of animal life: the hidden suffering and slaughter of animals on the factory farm literally makes the on-the-go meals available for the *Homo economicus* of biopower, today's busy lifestyle consumer.

Not surprisingly, Michel Foucault's work figures quite prominently in these emerging fields of study: Foucault of course coins the word *biopower* in *The History of Sexuality, Volume 1*. And in his lecture courses touching on the concept

(*Society Must Be Defended* and *The Birth of Biopolitics*) Foucault discusses the ways in which biopower might differ from the form of power he famously calls "discipline" (which aims at modifying individual behaviors and is always mediated through institutions). As Foucault explains in his 1975–76 lecture course *Society Must Be Defended*, biopower constitutes

> a new technology of power, but this time it is not disciplinary. This technology of power does not exclude the former, does not exclude disciplinary technology, but it does dovetail into it, integrate it, modify it to some extent, and above all, use it by sort of infiltrating it, embedding itself in existing disciplinary techniques. This new technique does not simply do away with the disciplinary technique, because it exists at a different level, on a different scale, and because it has a different bearing area, and makes use of very different instruments. Unlike discipline, which is addressed to bodies, the new non-disciplinary power is applied not to man-as-body but to the living man, to man-as-living-being.[2]

As Foucault insists, this new form of biopolitical power doesn't simply replace discipline but extends and intensifies the reach and scope of power's effects by freeing them from the disciplinary focus on the "exercise" and the "institution." Biopower, one might say, radically expands the scale of power's sway: by moving beyond discipline's "retail" emphasis on training individual bodies at linked institutional sites (family, school, army, factory, hospital), biopower enables an additional kind of "wholesale" saturation of power effects, smearing them across the entire social field. What Foucault calls this "different scale" and much larger "bearing area" for the practices of power makes it possible for biopower to produce more continuous effects, because one's whole life (one's identity, sexuality, diet, health) is saturated by power's effects, rather than power relying on particular training functions carried out in the discontinuous domain of X or Y institution (dealing with health in the clinic, diet at the supermarket and the farm, sexuality in the family and at the nightclub, and so on). Hence biopower works primarily to extend and intensify the reach of power's effects: not everyone has a shared disciplinary or institutional identity (as a soldier, mother, nurse, student, or politician), but everyone does have an investment in biopolitical categories like "sexuality," "health," or "quality of life"—our own, as well as our community's. If discipline forged an enabling link between subjective aptitude and docility,[3] biopower forges an analogous enabling link between the individual's life and the life of the socius: the only thing that we as biopolitical subjects have in common, one might say, is that we're all individuals, charged with the task of creating and maintaining our lives. And that power-saturated task is performed not solely at scattered institutional sites but virtually everywhere, all the time.

And the ethical challenge presented by animal studies arrives hot on the heels of a triumphant human biopolitics: most centrally, of course, there would seem to be serious concerns about the ethics of sacrificing animal life solely for human benefit (under a regime dominated by an intense concern for "life," why do we live and they die?), not to mention the sustainability consequences (both for individual health and the life of the ecosphere) that are being wrought by the huge corporate animal farms required to feed a growing global hunger for animal flesh.[4]

In any case it seems clear that Foucault's texts of the 1970s and 1980s constitute linchpin sites for both biopolitical analysis and the related fields that cluster under the rubric animal studies.[5] We should note, though, right from the beginning that Foucault seems to be of more use within contemporary biopolitical theory and practice than in the field of animal studies, where his work on biopower serves less as a common touchstone to build on than as a kind of negative jumping-off point—similar to the way that Theodor Adorno's supposed dismissal of popular culture served for many years as an enabling negative horizon for academic work in cultural studies.[6] In short, Foucault is routinely chastised for not paying enough (or really any) attention to the question of animality within his discussions of life as biopower. Most infamously, Foucault's work has been charged with "species chauvinism"—Donna Haraway's accusation in *When Species Meet*. Haraway, in fact, reveals that her book was spawned by the realization that Foucault's critical project didn't go far enough: "I had read Michel Foucault, and I knew all about biopower and the proliferative powers of biological discourses. . . . I had read *Birth of the Clinic* and *The History of Sexuality*, and I had written about the technobiopolitics of cyborgs. I felt I could not be surprised by anything. But I was wrong. Foucault's own species chauvinism had fooled me into forgetting that dogs too might live in the domains of technobiopower."[7] Perhaps, as I will suggest below, if she had spent more time with *The Order of Things* and *History of Madness*, things would have turned out differently.

Foucault is then given credit in animal studies for calling attention to the central question of life within modern political existence, but he's just as quickly disciplined for confining his analysis to humans, thereby doubling down on the nefarious ethical exclusion of animal life from the discussion. Nicole

Shukin writes in *Animal Capital*, for example: "The pivotal insight enabled by Foucault—that biopower augurs 'nothing less than the entry of life into history, that is, the entry of phenomena of the life of the human species into the order of knowledge and power'—bumps against its own internal limit at the species line. The biopolitical analyses he has inspired, in turn, are constrained by their reluctance to pursue power's effects beyond the production of the human social and/or species life and into the zoo-politics of animal capital."[8] There are undoubtedly a whole series of Foucauldian ways that one could respond to this kind of claim. I suppose the most obvious is that if Foucault is a booster for the human species—offering sunny thoughts like "man is an invention of recent date. And one perhaps nearing its end"[9]—I'd shudder to think what *critics* of the species might think. But rather than pursue this kind of defense (as Foucault himself insisted, polemical back-and-forth argumentation is unlikely to lead us "beyond" our present consensus),[10] I'd like to take a different tack. One way to go would be to suggest avenues whereby Foucault's work on human biopower could be harnessed for thinking about animal life. As Stephen Thierman writes in "Apparatuses of Animality: Foucault Goes to a Slaughterhouse," insofar as "Foucault did not write" extensively about nonhuman animals, perhaps "it is left to those of us who think his methods and conceptual tools can be fruitfully employed to explore our relationships with other animals to fill in the blanks."[11] And there's already plenty of work being done in this emergent area of research, analyses inspired by Foucault that interrogate the question of life across the human-animal divide.[12]

However productive that path might be, I'd like to take a somewhat different direction here, by trying to think about new trajectories in Foucauldian biopower not so much by extending his analyses into animal formations and institutions that he *didn't* study (contemporary corporate farming practices, genetic manipulation, the companion animal phenomenon, and so forth) but by looking at neglected formulations concerning animality and the emergence of biopower in his own work. I would begin simply by noting that Foucault hardly ignored animals altogether, especially in his early archaeological work. *History of Madness* (1961), for example, contains a substantial bestiary of reflection on the myriad historical ways that "the animal realm . . . serves to reveal the dark rage and sterile folly that lurks in the heart of mankind."[13] Foucault in fact insists that much of the discourse on madness, in the classical age and beyond, "took its face from the mask of the beast. . . . Madness . . . was for the classical age a direct relation between man and his animality, without reference to a be-

yond and without appeal" (*HM* 147–48). Human madness, as Foucault demonstrates, was for a very long time understood and treated as a kind of animality.

Even more centrally, an extended interrogation of animality figures in what
we might call the *first* birth of biopower within Foucault's corpus, in 1966's *The
Order of Things*. To begin reexamining the Foucauldian emergence of life as a
central concern for power, one needs merely to cite these controversial lines
from that work: "if biology was unknown [in eighteenth-century Europe],
there was a very simple reason for it: life itself did not exist. All that existed was
living beings, which were viewed through a grid of knowledge constituted by
natural history" (*OT* 127–28). Of course, this provocation—life did not exist in
Europe until the nineteenth century, specifically until 1802, when Lamark was
the first to use the word *biology*—functions as a bit of a dry run for Foucault's
later, seemingly just as "outrageous," idea that homosexuality was invented
in 1870, or his declarations that the author, and indeed even "man" itself, are
products of recent invention.[14] This type of sentence-level provocation is one
of Foucault's characteristic means of dramatizing, in a very stark way, the crucial importance of social and historical emergence. Alongside the Foucauldian
pleasure evident in the perversity performed by sentences like "up to the end
of the eighteenth century, in fact, life does not exist" (*OT* 160), there's a consistent philosophical point being advanced: the historical emergence of a new
way of handling topic X (here, "life") gives rise to different problematics, different practices, and thereby different objects. A new form of practice literally
remakes the (supposedly preexisting) object of the discourse: biology *creates*,
rather than *discovers*, this object of study called life. This emphasis on discursive emergence constitutes Foucauldian Archaeology 101 and keeps us focused
on the most basic terrain for all of Foucault's work: the question of how today
is different from yesterday.

In schematic terms Foucault's prey in *The Order of Things* is the epistemic
shift from a classical regime of representation (natural history, where classification and nomination of visible "living things" are the key practices) to a regime
of modern knowledge-transcendentals like life, labor, and language (where the
"object" of knowledge is no longer readily available to classification but rather
disappears into the shadowy half-light of discursive practice). In the birth of
biology the question of life unhinges itself from a practice of representation
(the discourse is freed from what Foucault calls the "pure tabulation of things"
[*OT* 131] in natural history's grids) and attaches itself instead to a mode of spec-
ulation about this murky thing called life—now understood *not* as a visible

manifestation of similitude but as the darkly hidden secret that connects living things. This movement from surface to depth signals the decline of natural history and the birth of biology, the emergence of the science of life. And in this movement across spheres, the modern human sciences and their era of transcendentals begin to replace the representational episteme—just as representation, in its turn, had replaced the early modern regime of fabulation or magic.

And it is here, in the interstices of this grand narrative about knowledge in the West, where animals make their appearance in *The Order of Things*: "To the Renaissance, the strangeness of animals was a spectacle: it was featured in fairs, in tournaments, in fictitious or real combats, in reconstitutions of legends in which the bestiary displayed its ageless fables. The natural history room and the garden, as created in the Classical period, replace the circular procession of the 'show' with the arrangement of a 'table,' . . . a new way of connecting things both [to] the eye and to discourse" (*OT* 131). As the historical a priori of representation emerges and later mutates into the era of the human sciences, the "being" of animals changes as well: in the era of representational natural history "the plant and the animal are not seen so much in their organic unity as by the visible patterning of their organs. They are hoofs and paws, fruits and flowers, before being respiratory systems or internal liquids. Natural history traverses an area of the visible . . . without any internal relation of subordination or organization" (*OT* 137). For a representational regime it is the visible surface of living things, rather than some buried "organic unity," that is the bearing area of discursive power. Natural history constitutes a series of practices whereby living things—plants and animals—find their proper classification through organization by common visible traits rather than hidden animating principles.

But on Foucault's account, it's with the rise of the transcendentals, in the era of the human sciences, that animals begin to take priority over plants as the privileged form or figure of life itself. In an era of natural history where knowledge was characterized by "the apparent simplicity of a description of the visible . . . the area common to words and things constituted a much more accommodating, a much less 'black' grid for plants than for animals" (*OT* 137). Most animals, simply put, have more hidden, interior space than plants and thereby present a greater volume of "black" or blank space to the gaze of the classifying naturalist. Foucault writes about this era of representation: "Because it was possible to know and to say only within a taxonomic area of visibility, the knowledge of plants was bound to prove more extensive than that of animals" (*OT* 137), precisely because plants can be pulled up out of the ground, and thereby rendered fully visible, from the tip of the roots to the outermost edges of the flower or leaf.

At the dawn of the nineteenth century, however, Foucault traces a mutation of the dominant epistemic procedures—from a representational discourse that maps external similitude and resemblance, to the emergence of a speculative discourse that takes as its object hidden internal processes. In short, we see emerge a discourse that "opposed historical knowledge of the visible to philosophical knowledge of the invisible" (*OT* 138): knowledge's privileged practices abandon the surface of objects in order to plumb their hidden depths instead. And first and foremost among those transcendental "invisibles" was a little thing we like to call "life": "The naturalist is the man concerned with the structure of the visible world and its denomination according to characters. Not with life" (*OT* 161), Foucault insists, because life is not representable. Life is in fact a kind of unplumbable depth, animating the organism from a hidden origin somewhere within. This birth of biology—which is to say, the emergence of "life" itself as a bearing area for discursive power and a depth to be explored—constitutes the first birth of biopower, this one in Foucault's work of the mid-1960s.

So, why is this archaeology of biopower important, in terms of using Foucault in the present and maybe leading us toward new directions in the future? Well, I suppose there are myriad answers to that question, but I'd like to suggest one particular line of useful inquiry here, one that also involves a way to respond to the animal studies critique of Foucauldian biopower with which I began this chapter—that Foucault essentially ignores the question of animal life and thereby extends the unearned privileges of human biopower rather than questioning them.

In short, Foucault's work on biopower 1.0 shows that animal life is not in fact jettisoned or abjected at the dawn of humanist biopower in the nineteenth century; instead, *animality is fully incorporated into biopower as the template for life itself.* As Foucault puts it,

the animal, whose great threat or radical strangeness had been left suspended and as it were disarmed at the end of the Renaissance, discovers fantastic new powers in the nineteenth century. In the interval, Classical nature had given precedence to vegetable values . . . with all its forms on display, from stem to seed, from root to fruit; with all its secrets made generously visible, the vegetable kingdom formed a pure and transparent object for thought as tabulation. But when the characters and structures are arranged in vertical steps toward life—

that sovereign vanishing point, indefinitely distant but constituent—then it is the animal that becomes the privileged form, with its hidden structures, its buried organs, so many invisible functions. . . . If living beings are a classification, the plant is best able to express its limpid essence; but if they are a manifestation of life, the animal is better equipped to make its enigma perceptible. (*OT* 277)

In short, Foucault argues that with the emergence of the human sciences at the birth of biopower, the animal is not excluded or forgotten but quite the opposite: animality is the dominant apparatus for investigating both what life is and what life does. The living is no longer primarily vegetable (sessile and awaiting mere categorization) but understood as evolving, appetite-driven, secret, discontinuous, mendacious, inscrutable, always on the prowl, looking for an opening to break free. As Foucault puts it, "Transferring its most secret essence from the vegetable to the animal kingdom, life has left the tabulated space of order and become wild once more" (*OT* 277). And this is of course not just a development within the narrow confines of biology. Foucault could in fact cue here the advent of philosophical modernity itself. One might speculate that Hegel's 1807 *Phenomenology of Spirit* (where human life itself is refashioned as nothing other than unfathomable discontinuity and animal appetite— in short, desire) shows the way for later nineteenth-century thought, which in turn opens a path directly to our day: from Darwin's evolution of life, through Freud's life of the unconscious, and Nietzsche's life of self-overcoming, all the way to Schumpeter's neoliberal life of creative destruction. All of these formations depend completely on the bedrock connection of life to an animating, hidden, "wild" animality of desire: both prior to and beyond the human yet somehow still constituting that humanity as its secret essence.

In *History of Madness* Foucault charts a similar archaeological shift in the role and status of animals within European discourse on the mad. As he notes, from the beginning "it was probably essential for Western culture to link its perception of madness to imaginary forms of the relation between men and animals" (*HM* 151), and for a very long time in the West animals played the role of uncontrolled counterhumanity—that crazed, menacing opposite to human reason: "animals were more often thought of as being part of what might be termed a counter-nature, a negativity that menaced the order of things and constantly threatened the wisdom of nature with its wild frenzy" (*HM* 151). "In the classical age," Foucault concludes, "madness was still thought of as the counter-natural violence of the animal world" (*HM* 151). However, further into in his *History of Madness* (foreshadowing the analysis in *The Order of Things*),

Foucault will note a decisive series of breaks in the Western discourse of madness and animality, at precisely the Enlightenment moment when madness begins to get incorporated into the definition of reason (as a potential malady or
even a secret romantic source of intellectual power, as opposed to the classical
formation wherein madness functions as a merely negative counterpoint to the
discourse of reason): "From the moment when philosophy became anthropology [when thought became focused on human life], and men decided to find
their place in the plenitude of the natural order, the animal lost that power of
negativity, and assumed the positive form of an evolution between the determinism of nature and the reason of man" (*HM* 151).

As in *The Order of Things*, Foucault shows us in *History of Madness* that
animality is not jettisoned at the birth of anthropological biopower (at least
in part because such an abjected binary otherness was the role of animality
in a prior, classical era); rather, "when philosophy became anthropology," the
animal became incorporated into reason. Animality then becomes less a binary
all-or-nothing function in relation to madness, operating instead on a sliding
scale. With the anthropological turn to life the madman is recast: not as the
subhuman animal other but as our less-fortunate sibling—beset not by the classical era's understanding of animality as the absence of rationality but by too
much of a good thing: biopower's embrace of life as animal desire.

Likewise, in *History of Madness* Foucault also calls our attention to the Aristotelian definitions of man (as rational and political animal) that will play a
huge role in his thinking about biopower in the 1970s and 1980s:

> Did the fact that, after Aristotle, men had spent two thousand years thinking
> of themselves as reasonable animals necessarily imply that they accepted the
> possibility that reason and animality were of a common order? Or that the defi-
> nition of man as a "rational animal" provided a blueprint for understanding
> man's place in natural positivity? Independently of whatever Aristotle meant
> with that definition, it might be the case that for the Western world "rational
> animal" meant the manner in which the freedom of reason took off from a space
> of unchained reason and marked itself off from it, ultimately forming its op-
> posite. (*HM* 151)

But as philosophy becomes anthropology at the birth of biopower in the early
nineteenth century in Europe, Foucault argues that the classical age of the animal's binary alterity is also eclipsed: "At that point, the meaning of the term
'rational animal' underwent a radical change. . . . From then on, madness had

to follow the determinism of a humanity perceived as natural in its own animality" (*HM* 151).

In *History of Madness* Foucault suggests we look to "the French poet Lautré-
amont" for "proof" of the "wild frenzy" (*HM* 151) that is life-as-animality; but
consider, just as a passing example nearer to the present, Frank O'Hara's 1950
poem "Animals":

> Have you forgotten what we were like then
> when we were still first rate
> and the day came fat with an apple in its mouth
>
> it's no use worrying about Time
> but we did have a few tricks up our sleeves
> and turned some sharp corners
>
> the whole pasture looked like our meal
> we didn't need speedometers
> we could manage cocktails out of ice and water
>
> I wouldn't want to be faster
> or greener than now if you were with me O you
> were the best of all my days[15]

Here in O'Hara's poem, as Foucault suggests in the larger biopolitical realm of
modernity, animals function less as our excluded "other" than as very intense
markers for our hidden, better, or former—perhaps more authentic—selves.
Our unconscious drives, O'Hara suggests, are animal in nature, and it is those
unbridled desires that make us "first rate," most truly who we are. As O'Hara
puts it, "the best of all my days" were animal in nature—when I didn't worry
about Time, traffic, or stocking the bar, and "the whole pasture looked like our
meal." Here, animality is not demented or predatory (as Foucault suggests the
classical age understood the mad, irrational "otherness" of animals) but takes
the form of a biopolitical or "more gentle form of animality, which did not de-
stroy its human truth in violence but allowed instead one of nature's secrets to
emerge: the rediscovery of the familiar but forgotten resemblance [of the mad
person] with tame animals and children" (*HM* 435). I take this to resonate with
O'Hara's conclusion, where it is the precocious, desiring animal in us all who is
addressed in the final lines: "O you / were the best of all my days."

In short, the archaeology of biopower that Foucault performs in *The Order
of Things* and *History of Madness* shows decisively that the jettisoned, negative,

or forgotten other of our biopolitical conception of life is most assuredly *not* the animal, insofar as animality is the subtending paradigm for our era of humanist, neoliberal biopower—where it's all appetite and appropriation all the time.[16] Tweaked and intensified a bit, Foucault's archaeology of animality may compel us to ask whether contemporary animal studies, far from constituting a critique of an all-too-humanist biopower (exposing the imperialism of human life over animal life), tends to function in fact as an intense extension of that very biopower. A Foucauldian provocation might suggest that animals are important within contemporary academic discourse, or at least they're more important than plants or other forms of life, not because animals function ethically as "wholly other" to humans but primarily because they're "like us" in an originary way: they experience intense feelings; they are born and they die; they like to go walking in the park at sunset. Indeed, animals are wild, just like we are in our best and freest moments. As such, though, animals function less as our ethical "absolute other" than as our hidden, better self: our unconscious drives are animal in nature, and it is those unbridled desires that make us most truly who we are. Animals are more our life "companions" (to steal another phrase from Haraway) than our "others" (those figures excluded, forgotten, wholly unlike us but that we still depend on absolutely). Following Foucault's reading, one might suggest that role of abjected other as having been played throughout the biopolitical era not by the animal but by the plant—which was indeed forgotten as the privileged form of life at the dawn of biopower. In this context it is probably worth recalling that the biomass of plant life on Earth's terra firma does remain approximately one thousand times greater than the combined zoomass of all humans and other animals.[17]

And perhaps one related upshot of this genealogy, in terms of new directions in biopolitical ethics, might be an imperative to look at the strange and consistent elision of plants within the voluminous work on life within contemporary theory and philosophy—the primary project that will engage me in this book. There are, as we will see, myriad tantalizing places in Heidegger, Agamben, and Derrida (all of them rereading Aristotle, especially from book 2 of *De anima*) where the proximity and difficulty posed by plant life is highlighted, only to be dropped quickly and consistently by all these thinkers in order to stay on the trail of the human/animal distinction. Indeed, if animal studies scholars can charge Foucault with "speciesism," then in turn Foucault's archaeology of life (from the privileged plant form in an earlier era to the biopower's fetishizing of the animal) might suggest that animal studies, in its foundational abjection of plant life, is guilty of "kingdomism"—ignoring not just a species but an entire kingdom, which one would assume is a much greater crime on the "kingdom-phylum-class" sliding scale of differentiation. Likewise, continuing research into plant behavior and intelligence confirms that plants are not, as was thought for centuries following Plato and Aristotle, sessile and insentient: recent research has uncovered that plants evidence active, purposeful, future-oriented movement and exhibit both competitive and defensive behavior. Plants, it seems, also have a certain kind of language—they share information concerning soil conditions and the presence of predators.[18] Given this series of revelations, it becomes even harder to draw the lines among human, animal, and vegetable life. Indeed, some recent botanical research even bolsters Aristotle's twenty-four-hundred-year-old quip about feeling plants' pain: "A plant which is fixed in the ground does not like to be separated from it."[19]

As I noted in my Preface, there are already several scholarly works out there that interrogate the question of vegetative life. Most of these books are notable, for my purposes here, not so much because they attempt to question the (animal) life-as-hidden-secret model that Foucault diagnoses for us but because they work very hard to extend that "hidden life" paradigm (and the anthropomorphic identity logic that rules over it) to plants as well.[20] The "secret life of plants" has been a leitmotif from the dawn of interest in this topic in the 1970s, all the way to a 2013 symposium with that same title at Princeton.[21] Mining this secretive nature, Marder suggests that "to get in touch with the existence of plants one must acquire a taste for the concealed and the withdrawn, including the various meanings of this existence that are equally elusive and inexhaustible." He continues: "if we are to 'think the plants,' we must not shy away from darkness and obscurity," and "the generosity of vegetal soul is inexhaustible."[22] In short, a book like Marder's *Plant-Thinking* will argue at the end of the day that plants are the new animals (in the sense that Lauren Berlant conjures when she suggested that "affect is the new trauma").[23] And there is likewise a robust emergent literature in both environmental studies and machine ethics, which suggests, if nothing else, that the question "What counts as an ethically compelling form of 'life'?" is and will remain an open and hazardous one.

In any case, as Derrida persuasively insists in his work on animals, the primary stake of interrogating animality is not asking what we humans ethically need to "grant" to animals (personhood, thinking, recognition, a voice, a face, and so on), or to treat animals as identical to humans, but in asking whether humans can somehow separate themselves from (or elevate themselves above) their conception of "subhuman" others, like animals. (As we will see in the coming chapters, the Derridean project is undertaken not in the name of granting human privileges like rationality, communication, and agency to animals but in wondering whether humans have any less fettered access to those things than animals do.) And the conundrum would be ethically similar, it seems to me, for whatever we might have to say in the future about plants or other forms of biopolitical life: the project is less offering some of our human privilege to plants or machines or the earth itself (in short, anthropomorphizing them) than paying close attention to the power effects rendered by the myriad practices by which we do in fact differentiate ourselves from other forms of life, and what forms of violence those practices inevitably inflict.

Foucault of course parts ethical company with Derrida (and, I would suggest, with the founding principles of much animal studies) around the binary pathos of "totalization or nontotalization," which constitutes nearly the whole field of ethics in a deconstructive context: if totalization or the violent desire for completion can be disrupted, if an originary différance of undecidability can be mobilized and demonstrated, then some positive deconstructive work has been accomplished. However, such a supposedly ethical gesture toward the unfathomable or untotalizable other, as Foucault insists throughout his work, poses no essential question (ethical or otherwise) to the human sciences, because those contemporary sciences don't require or even desire totalization. As Foucault demonstrates in his work on the emergence of life in Europe, the Western human sciences need constantly to refashion an unfathomable depth, and inexhaustible other, so they can continue to do their work. The insistence on the primacy of some nontotalizable other doesn't cripple the human sciences but rather constitutes almost the entirety of their work: as Foucault concisely puts it, "an unveiling of the non-conscious is constitutive of all the sciences of man" (*OT* 364). (Economics, for example, doesn't know what value is any more than theology knows what God is or biology knows what life is—that's why you have robust discourses to study them.) So the trading-places game of ethical alterity—the nonhuman other is best figured as the unconscious, the animal, the plant, the earth, the robot, and so forth—tends primarily to extend and deepen the constitutive work of the human sciences (the production of undecidability, which in turn produces more commentary), rather than to disrupt that work in some essential way.

In fact, barely a page of Foucault's methodological treatise *The Archaeology of Knowledge* goes by without some kind of stinging critique of any and all discourses of absent origin, hidden depth, or undecidability. As Foucault clearly writes, what he seeks in archaeology is "not a condition of possibility but a law of coexistence": discourse is constituted by "something more than a series of traces" of lost origins, and he insists that the statements forming and transforming the archive are "not defined by their truth—that is, not gauged by the presence of a secret content."[24] In his most pointed criticism of Derrida in the *Archaeology*, Foucault questions the manner in which discourse "can be purified in the problematic of trace, which, prior to all speech, is the opening of inscription, the gap of deferred time [*écart du temps différeré*]: it is always the historico-transcendental theme that is reinvested" (*AK* 121). And, mirroring language that he uses to criticize Derrida's reading of *History of Madness*,[25] Foucault lays waste to any discourse, scientific or philosophical, "which finds, beneath events, another, more serious, more sober, more secret, more fundamental history, closer to the origin, more firmly linked to its ultimate horizon (and consequently more in control of all its determinations)" (*AK* 121). In the end, for Foucault one might say that deconstruction is merely a marker and bearer of the "nontotalizing" *symptoms* of biopower and the human sciences, rather than forwarding any kind of corrective to these formations.[26]

Foucault and Agamben Redux: Animality and Biopower

I think one can see more clearly Foucault's difficulties with the whole transcendentalist discourse of totalization and hidden connection by looking at Giorgio Agamben's reading of him. Agamben (in)famously outlines his project in *Homo Sacer: Sovereign Power and Bare Life* like this: "the Foucauldian thesis [on biopower] will . . . have to be corrected or, at least, completed."[27] Agamben gives Foucault credit for calling our attention to something like the move I have just highlighted in Foucault's work, the biopolitical subsumption of life's animality (Agamben's *zoe*) within the realm of social power (*bios*): "In the last years of his life . . . Michel Foucault began to direct his inquiries with increasing insistence toward the study of what he defined as biopolitics, that is, the growing inclusion of man's natural life in the mechanisms and calculations of power" (*HS* 119). And certainly Agamben's work on the "exception" follows a logic somewhat similar to Foucault's on the "norm": alterity is incorporated precisely by being designated as other; the abnormal exception reinforces the power of the norm. However, Agamben's completion of Foucault functions not through a series of minor corrections but rather through a wholesale rejection of Foucault's privileged biopolitical sites of historical analysis—sexuality, market economics, or the fashioning of the self—in favor of an emphasis on Nazi concentration camps as the "exemplary places of modern biopolitics" (*HS* 119); as we will see, that change of diagnostic venue makes all the difference when it comes to thinking life and animality.

As Agamben insists, "today [in 1995] it is not the city but rather the [concentration] camp that is the fundamental biopolitical paradigm of the West" (*HS* 181), by which he primarily means to signal the continued primacy of sovereignty as the contemporary mode of political power, specifically in the form of the sovereign exception: who lives and who dies; those granted inclusion and the life of citizenship versus those who are "othered" and reduced to bare life. "At once excluding bare life from and capturing it within the political order, the state of exception actually constituted, in its very separateness, the hidden foundation on which the entire political system rested" (*HS* 9). In a move quite familiar to us from deconstruction, Agamben shows us how the abjecting of "the other" (homo sacer) is in fact the condition of possibility for the configuration of "the same" (the citizen). The very biopolitical exclusion of zoe (bare life) allows bios (political life) to emerge; zoe is thereby incorporated into bios as that which is excluded but can never be acknowledged as such, the exception that proves the sovereign rule.

While Agamben consistently acknowledges his debts to Foucault in all this (as he says, "I first began to understand the figure of the *Homo sacer* after I read Foucault's texts on biopolitics"),[28] there seems a long line indeed of things that Agamben means to "correct" in Foucault. As Derrida scathingly puts it in his treatment of Agamben, "poor Foucault—he never had such a cruel admirer."[29] So what exactly is it that's being "corrected or, at least, completed" by Agamben? It seems to me that the stake is nothing less than Foucault's entire analysis of power. Recall first the Foucauldian axiom that each historical mode of Western power dominant since the sovereign era of the early modern monarchs (the disciplinary regimes that first arose in the eighteenth century and the biopower born in the nineteenth) has not primarily been characterized by a top-down state apparatus (much less by the sovereign decisions of particular individuals) but works at more mundane "retail" levels. Foucault consistently emphasizes the ways in which power over the past several hundred years has become "intensified"—disbursed and saturated within institutions (discipline) and even into the everyday work of subject formation itself—in practices like sexuality or the neoliberal emphasis on an individual consuming market goods and services (under the dispositif of biopower).

In Foucault's account Western "power" has not operated primarily in a "sovereign" manner (centralized in a very small set of totalitarian decision makers bent on pure domination) since the seventeenth century—which is not to say that sovereign power has disappeared altogether but that it is no longer the primary mode through which other modes must make their way. Sovereign power still exists under disciplinary and biopolitical regimes, but to do its work, even this top-down brand of sovereign power needs to pursue its aims through other-than-sovereign means—through institutional channels (for discipline) or subjective ones (biopower). Most important, the rise of capitalism (and thereby the dismantling of state-based sovereignty) figures prominently in Foucault's account of biopower's intensification; and in this Foucault's analysis is akin to Marx's, where the "real subsumption" of capital is impossible until capitalism morphs from being a kind of external shell containing the socius (merely "formal subsumption") to a state where capitalism becomes completely woven into the fabric of everyday life.[30]

For Agamben, on the contrary, it is not decentralized capitalist practice but the sovereign state of exception—a centralized, top-down *decision* concerning who lives and who dies—that, far from constituting an archaic mode or understanding of power, remains the essence or hidden animating secret of power in the West, and to this day it constitutes the basis for all political power. In short, for Agamben nothing short of what he names "totalitarianism"—the idealist project of wholly eradicating difference and resistance—is the political project characteristic of power in what he calls "our age." In correcting Hannah Arendt's notion of power along the path of tidying up Foucault's mistakes, Agamben writes that "the radical transformation of politics into the realm of bare life (that is, into the camp) legitimated and necessitated total domination" (*HS* 120): "the [concentration] camp—as the pure, absolute, and impassible biopolitical space (insofar as it is founded solely on the state of exception)—will appear as the hidden paradigm of the political space of modernity" (*HS* 123). For Agamben the machine that animates all the others today is neither a disciplinary institution nor a sexual identity but a Nazi concentration camp—a "pure, absolute and impassible biopolitical space" dedicated to "total domination." Right now, update your Facebook status to "totally dominated."

Of course, it's precisely this melodramatic strain that makes Agamben so seductive for many readers, but from a Foucauldian point of view it seems a serious diagnostic mistake to suggest that the biopower characteristic of global neoliberal capitalism does its work or achieves its effects in the same way as the state-based fascist biopower of the early to mid-twentieth century did. Aside from a very loose metaphorics, I see no convincing way that the "dream structures" of post-postmodern capitalism (the online auction or dating site, the shopping mall, or the stock market) are "fascist" or "totalitarian" in the same way that a concentration camp was, and insisting on such a homology seems to offer very few tools for understanding, much less resisting, the supple biopower of our own day. Capital requires us all to circulate, as much and as widely as possible—looking for work, bargains, love, information, Internet porn, recipes, entertainment, new experiences, whatever. Inversely, the concentration camp exists primarily to curtail all movement; it's a machine designed to confine and warehouse human subjects until it's time to slaughter them. Why contemporary consumption capitalism wants to confine and then slaughter its primary drivers (consumers) is never quite made clear in Agamben. In any case there would seem to be important differences between the Luftwaffe and DirectTV, though both can overwhelm you, unseen from the air. At the very least it would seem that the parallels among market economics, culture-industry manipulation, and the techniques of fascism that Adorno identified in the mid-twentieth century would have to be updated considerably rather than simply imported wholesale into our present.

As I've pointed out elsewhere, the suturing of contemporary subjectivity and quotidian life to the regime of biopower (which is at work everywhere, from sexuality through neoliberal consumerism to the everyday work of each person's identity formation) constitutes for Foucault a "postulate of absolute optimism" (as opposed to the unrelentingly tragic pathos of Agamben's work on the regime of biopolitics).[31] It's axiomatic for Foucault that where there is power, there is resistance; so as biopower saturates every corner of our daily lives, so do the experimental practices of resistance: more saturation of power also means more sites of resistance. As our everyday lives become increasingly targeted by biopower, so do our everyday practices themselves become sites of increasingly intense resistance. Foucault's work demonstrates time and again that power does not primarily dominate or repress in some totalitarian fashion; in fact, Foucault famously argues that almost all political theory has yet to cut off the king's head—which is to say, political theory remains ineffectual because of its outmoded understanding of how power works. He writes quite straightforwardly: "What we need . . . is a political philosophy that isn't erected around the problem of sovereignty."[32] In short, Foucault's provocation concerning power is simply this: we are freer than we think we are.

For Agamben nothing could be farther from the case: as he argues, the "increasing inscription of individuals' lives within the state order" of biopolitics inexorably leads to "a new and dreadful foundation for the very sovereign power from which they wanted to liberate themselves" (*HS* 121). In Agamben contemporary biopower is and remains sovereign in practice and theory. So the whole of Agamben's project, one might say, consists of stitching the king's head back on, recentering and hierarchizing power—thereby eschewing virtually the entirety of Foucault's work on biopower, not to mention his analysis of the disciplinary regimes, in favor of what in Foucault remains a seventeenth-century conception of power. Odder still is the historical fact that Agamben's work in this area has taken place during an unprecedented period of neoliberal privatization of social power, since the 1990s. Entire populations today undoubtedly continue to find themselves newly categorized as expendable "bare life"; however, the practices responsible for that categorization are increasingly implemented not by jackbooted state thugs who round up and confine populations at state-run facilities but by neoliberal market-makers in designer suits (who, make no mistake, have access to thugs should it come to that; but the thugs aren't the cutting financial edge of the operation: not efficient, too expensive).

Corporate risk assessment teams and credit rating agencies redline existing neighborhoods (or even entire countries, as Greece or Ireland could attest after the 2008 worldwide crash); and concomitant "austerity" measures mean less access to jobs, affordable food, and health care. But such measures are dictated and brought about primarily through market mechanisms; and while large-scale austerity plans continue to be installed through sovereign dictates of the nation-state—protecting the currency and servicing debtors before citizens, which means slashing social benefits—those interventions are largely undertaken today in the name (which is to say, within the practice and theory) of the market, not in the name of the *Volk*, the Supreme Leader, or the king. There are of course no legal barriers against a corporation opening a discount supermarket or a health-care facility in a "high-risk" neighborhood, but there are no market "incentives" to do so either; and these biopolitical disincentives function more effectively than the former state-based modes of discrimination. (Racial desegregation, women's or ethnic minority rights, equal access laws concerning physical disability or sexual orientation: these are all protected by the law, but continue to be exercised—or not—largely by the market.) But of course the punch line, then, is that, given this state of affairs, the dictates of the market are anything but states of exception; they are the (ubiquitous) rule. Or, more

accurately, market mantras like flexibility and resilience are themselves made
into rules, ones that extend everywhere (the family, the school, the hospital, and
so on). Indeed, neoliberalism's stated desire to make government "small enough
to drown in the bathtub" (in Grover Norquist's pithy formulation) ensures that
even if any given state's interests in socialism somehow should rekindle, there
will be little or no tax revenue to support such ventures. In any case this unprecedented global rise of neoliberal, antigovernment sentiment and practice over
the past several decades seems at considerable odds with Agamben's portrait of
state power's near-iron centralization "today." The concentration camp, whatever
else one might say about its unspeakable horrors, was most certainly not driven
by the dictates of today's cowboy-capitalist, "small government" neoliberalism.[33]

As Foucault famously puts it, today's dispositif of biopower works by "making live and letting die," as opposed to sovereignty's rounding up the usual sus-
pects and killing them (or not)—"making die and letting live." And this is not,
as Agamben would have it, simply an update of sovereign power's relation to
"life" but a wholesale transmutation of it: Foucault's biopower works directly
on life, from the very beginning (birth rates, demographics, infant health, im-
munizations), in contrast to sovereign power, which is only really interested in
most people's everyday life in a negative or external way (extensive actuarial
knowledge concerning the everyday sex lives and health records of peasants
is inconsequential to the early modern sovereign but have become key to the
constant risk assessments of neoliberal capitalism). As Gilles Deleuze argues in
his essay periodizing Foucault's work (from discipline to biopower), "a man is
no longer a man confined but a man in debt"[34]—which is maybe to say that we
are "made to live" in a biopolitical register primarily by being encouraged to
consume (follow our desires), sometimes far beyond our means.

In Agamben's work, however, contemporary biopolitical questions of debt,
surveillance, or the increasing saturation of the culture industries (questions
concerning a control that steps in to intensify the disciplinary mechanisms of
the panopticon) are finally just more evidence of totalitarian sovereignty, bent
on reducing us to bare life through the sovereign exception. Agamben sums
things up in "No to Biopolitical Tattooing," his 2004 explanation of why he
would not submit to post-9/11 US customs fingerprinting:

> The problem exceeds the limits of personal sensitivity and simply concerns the
> juridical-political status (it would be simpler, perhaps, to say bio-political) of cit-
> izens of the so-called democratic states where we live. There has been an attempt
> the last few years to convince us to accept as the humane and normal dimensions

of our existence, practices of control that had always been properly considered inhumane and exceptional. Thus, no one is unaware that the control exercised by the state through the usage of electronic devices, such as credit cards or cell phones, has reached previously unimaginable levels. All the same, it wouldn't be possible to cross certain thresholds in the control and manipulation of bodies without entering a new bio-political era, without going one step further in what Michel Foucault called the progressive animalisation of man which is established through the most sophisticated techniques.[35]

This sense of "progressive animalization of man which is established through the most sophisticated techniques" is probably the most concise version of what Agamben wants from Foucault, or what attracts him initially to what he sees as Foucault's analysis of biopower: quite simply, the more bios we endure (the more our lives are subject to increasingly sophisticated political media and demographic techniques for working on individuals), the more animalized and expendable we become (the more we are reduced to the bare, bestial life of zoe).

Indeed, if we examine Agamben's earlier deployment of this Foucauldian turn of phrase concerning biopower and "animalization" in the opening pages of *Homo Sacer*, it's clear that Agamben wants from Foucault the outlines of a tragic modernist logic—where progress leads inexorably to barbarism, refining life means eliminating life, bios becomes indistinguishable from zoe, human life is reduced to animality. In a kind of "pure gold" moment for Agamben this Foucauldian sentiment concerning the "animalization of man" is followed directly by one of Foucault's very rare mentions of the Holocaust (in some remarks after a paper given at Stanford in 1979). Very early on in *Homo Sacer*, Agamben outlines and bootstraps his project by quoting from Foucault:

> After 1977, the courses at Collège de France start to focus on the passage from the "territorial State" to the "State of population" and on the resulting increase in importance of the nation's health and biological life as a problem of sovereign power, which is then gradually transformed into a "government of men." [Here Agamben cites Foucault's collected *Dits et écrits* 3:719, where this quotation does live; the following one does not.] "What follows is a kind of bestialization of man achieved through the most sophisticated political techniques. For the first time in history, the possibilities of the social sciences are made known, and at once it becomes possible both to protect life and to authorize a holocaust." (*HS* 3)[36]

The logic that Agamben wants to harvest from Foucault could not be clearer: the ascent of biopower—the increasing "government of men" and their biological lives—results directly in mass slaughter: the political project of protecting and managing a healthy life (bios) for the citizenry simultaneously makes it necessary to exclude and eliminate other, inferior or dangerous forms of life (zoe). Biopower is the operating system of totalitarianism, and modern power is in its nature sovereign, with its primary tool kit being the permanent state of exception (the decision concerning who lives, who dies) and the constant related threat of being reduced to bare, animal life. In short, Agamben extracts from Foucault a biopolitical logic whereby the human sciences (the study and maintenance of life) and the Holocaust (life's obliteration before power) constitute a Mobius strip of political rule. To "protect life" is in the end also and necessarily "to authorize a holocaust"; our haughty "sophisticated political techniques" (bios) in the end only reduce us to beasts (zoe), to be controlled by our handlers until they decide to slaughter us like so many cattle.

All this is fine and good, the sort of tragic fare that makes Agamben such appealing reading for academic liberals—good news for people who like bad news. But as I've been suggesting, whatever you may think of Agamben's diagnosis of the contemporary (and it should be clear that I'm not persuaded), you simply can't get there from Foucault. And there are really no reasons why you'd want to, insofar as the Foucauldian biopolitical world of bodies, pleasures, and a thousand microresistances are of little to no use in a totalitarian society modeled on the concentration camp, where you are forced to obey or be obliterated. If you're reduced to animality by power, your resistances can't really count for much, or any more than the steer's thrashing moments of protest before its death constitute any meaningful resistance to sovereign power of the slaughterhouse.

Strictly speaking, however, the horrific human-slaughterhouse situation of a concentration camp is not a Foucauldian "power" relation at all, simply because in Foucault a power relation implies resistance, which is by definition absent in the situation of "bare life" or pure domination. If Agamben is right—if power today remains sovereign in its nature and practice—then Foucault's work on biopower and the increasing ubiquity of resistance becomes completely moot. This is precisely why Agamben has so little use for Foucault's everyday sites of analysis: "Like the concepts of sex and sexuality, the concept of the 'body' too is always already caught in a deployment of power. The 'body' is always already a biopolitical body and bare life, and nothing in it or in the economy of its pleasures seems to allow us to find solid ground on which to oppose the demands of sovereign power" (*HS* 187). If we are all akin to Damiens, the tortured, animalized regicide who endures the intensity of sovereign power in the horrific opening pages of Foucault's *Discipline and Punish*, then I think we'd have to agree that Agamben is right: our sexuality or our everyday responses to the practices of power do indeed remain completely unhelpful when it comes to mobilizing resistance. Any more than a barnyard rooster's courting of a hen, it's hard to imagine Damiens's sexuality or other of his microlevel practices of everyday resistance having put up any bulwark at all against his reduction to a pure surface of bare life, to be eliminated (or not) by sovereign power. If we are all Damienses—reduced to animality (made to die or let live) by our sovereign masters—perhaps our only solace for the future is that "We'll Make Great Pets," as Perry Farrell has ironically encapsulated the fate of the human.

To put it another way, if the "progress" of the human sciences has an obvious relation to the "animalization" of "man," I've tried to show in what sense this functions in Foucault: biopower "animalizes" not through a chiasmic reversal or "reduction" of the human to its supposed other, "bare" "animal" life (through Agamben's biopolitical tattooing). Rather, biopower "animalizes" in Foucault through a subtending incorporation of an understanding of "life" primarily as desire (from the representational era's picture of life as a kind of sessile plant, to the roaming, restless animal who becomes the incorporated figure for biopower). Insofar as the human animal is understood in terms of its desires under the regime of biopower, it certainly follows for Foucault that those desires can be worked on by sophisticated political techniques—indeed, compared to the relatively crude techniques of sovereign power (torture and public execution), biopower needs to build on disciplinary exercises by inventing ever-more supple techniques. Insofar as nonsovereign power acts on actions (rather than on bodies or minds), it must become "light" enough to regulate virtual states of practice. The sophistication of biopolitical techniques, in other words, is not for Foucault primarily to be found in their brutish totalized or totalitarian intentions (power's hidden desire to reduce us all to beasts, leashed and branded by power) but in those techniques' lightness, their increasing saturation within the socius and their heightened economic effectiveness, what I've elsewhere highlighted as Foucauldian power's "intensity."[37]

When Foucault insists that there's an "animalization of man" involved in biopower's birth and functioning, he means it quite literally: we have incorporated the beast into the contemporary biopolitical definition of "man" as endless, unthematizable animal desire, with the practices of sexuality and neoliberal capitalism its two most intense markers. Perhaps this explains why Foucault good-naturedly tells Deleuze, "I cannot bear the word desire,"[38] because for Foucault the theme of desire quite literally continues to play out a tired dialectical drama born at the dawn of the nineteenth century: animal desires endlessly frustrated by the constraints of civilization.

For Agamben, however, bestialization constitutes less a contemporary practice or a historical phenomenon than a transhistorical metaphor or simile for the human condition, as is (despite his protests to the contrary) his emphasis on the concentration camp or sovereign power. Contemporary society is *like* a concentration camp or *like* an absolute monarchy; we are treated *like* animals when we have to surrender our DNA or fingerprints. But if animal studies has taught us nothing else, its emphasis on the material facts of the food industry should make us suspicious of metaphors suggesting that humans are treated as animals by advanced capitalism: recall that more than fifty billion chickens are slaughtered each year globally (more than nine billion in the United States alone), the vast majority living their short, genetically engineered lives in a cage that's too small for them to stand up or turn around in, many subsisting on feed made from the ground-up corpses of their deceased brethren, too unfit for sale. As utterly terrible as global poverty is to endure for half the human population on this planet, most people are not in fact treated like animals—slaughtered by the billions in what is indeed a sovereign manner, with little or no afterthought.

Indeed, as Foucault puts it in his formula for biopower, today's global poor are made to live, or left to die—an important difference from the sovereign, concentration-camp edicts of making die, or letting live (which most animals presently endure: the feed animals are made to die, while the dwindling populations of "wild" animals are merely left to live). Of course, ethnic cleansing, torture, and mass slaughter of humans (Foucault's sovereign "making die") still does happen in our world, but those sovereign practices are hardly the primary modality of power's functioning today (small solace, however, if you happen to be subject to these horrific practices). Or at least this is Foucault's bedrock argument about power's mutations in the West, pivoting over the last four hundred years from the sovereign execution of Damiens to a focus on the biopolitcal intricacies of our identity, health, sexuality, or consumer desires.

And this diagnostic difference between sovereignty and biopower is crucial at least partially because sovereign power, while notoriously difficult (if not impossible) to resist, tends to be relatively easy to spot, diagnose, and denounce: in short, someone else is always wielding "sovereign power." Even so, the biopolitics of "making live and letting die" is a regime in which all of us are implicated: who gets antiretroviral drugs, and who doesn't? Are the rich countries willing to pay more taxes, or endure weaker corporate profits, so that millions of poor people in remote regions can live? Are such policy matters reducible to a model of sovereign "decision"? The dictates of biopower implicate us all in the global dramas of life and death (it's a neoliberal market question all the way down), in a way that the bare life / sovereign power drama doesn't. As Tim Dean puts it:

> In his attempt to revise Foucault, Agamben instead reverts to a pre-Foucauldian model that treats power relations as polarized between those who have all the power (the position of sovereignty) and those who have none (the position of homo sacer). . . . I suspect that the appeal of Agamben's account, while ostensibly attributable to its utility in describing our post-9/11 political landscape, lies more fundamentally in its reassurance that we know where the power is because it has been so starkly consolidated. *Homo Sacer* permits us once more to believe in, and perhaps identify with, the reassuring idea of wholly innocent victims, those who have been divested utterly of power.[39]

In short, Agamben's tragic thematization of the world is utterly irreconcilable with Foucault's, precisely around the central Foucauldian question of what power is and how it operates. For Foucault, biopower does not primarily function in a "sovereign" manner, and to suggest that it does violates the cardinal, organizing rule that is consistently repeated throughout his work—the simple lesson that power is neither good nor bad but dangerous. It's not held by some and withheld from others, but it circulates through a socius. For Agamben, however, political power today remains sovereign (it either takes life or simply lets live, even in its most "sophisticated" forms), and thereby power, as Dean insists in his account of Agamben, is understood as being very bad indeed.

Zoe and/as Vegetable Life?

This largely metaphorical drama of sovereign power versus bare life helps us to understand why, in *The Open: Man and Animal*, Agamben can argue not only that the animal is a figure for mere or natural life (zoe) but, indeed, that vegetable life is or can be one of its primary figures. For example, Agamben somewhat puzzlingly writes on the heels of his analysis of Bichat:

> As Foucault has shown, when the modern State, starting in the seventeenth century, began to include the care of the population's life as one of its essential tasks, thus transforming its politics into biopolitics, it was primarily a means of a pro-

gressive generalization and redefinition of the concept of *vegetative life* (now coinciding with the biological heritage of the nation) that the State would carry out its new vocation. And still today, in discussions about the definition *ex lege* of the criteria for clinical death, it is a further identification of this *bare life*—detached from any brain activity and, so to speak, from any subject—which decides whether a certain body can be considered alive or must be abandoned to the extreme vicissitude of transplantation."[40]

Here Agamben startlingly suggests that his notion of bare life is in fact best understood not as "wild" animal life but as vegetable life, the lowest form of the living within Aristotle's tripartite vegetable-animal-human work on the *psukhe* (which we will examine in Chapter 2). Even more puzzling here is the claim that Foucault grounds his analysis of state racism ("the biological heritage of the nation") in "the concept of vegetative life." That's an interesting claim, one that I'd like to hear more about.

Rather than further develop these startling claims, however, Agamben immediately backs off this tangent about vegetable life as bare life (this in fact will be the last mention of vegetable life in the entire book), returning to the familiar ground of bios, the political life of humans, and zoe as the animal "bestialization" of that life:

> The division of life into vegetal and relational, organic and animal, animal and human, therefore passes first of all as a mobile border within living man, and without this intimate caesura the very decision of what is human and what is not would probably not be possible. It is possible to oppose man to other living things, and at the same time to organize the complex—and not always edifying—economy of relations between men and animals, only because something like animal life has been separated within man, only because his distance and proximity to the animal have been recognized first of all in the closest and most intimate place.[41]

So from the bare life of vegetality, seemingly the threshold between life and nonlife (from Aristotle forward, it's held that plants are "alive" in a way that rocks are not), we withdraw somehow to an even more originary threshold between man and animal—suggesting that for Agamben, Aristotle's tripartite soul (vegetable-animal-human) grows not from the ground up but from the top down. This is, I think, only to emphasize the obvious: that everything in Agamben goes through the bios/zoe distinction, with animality constituting the transhistorical other (the first sovereign exclusion) that haunts and disrupts all "civilized" or cultured human life. In short, the zoe of animality becomes "the open" from which the bios of human political life must consistently wrest itself but to which it just as consistently has to appeal. And any other conceivable or real form of life thereby becomes subordinated to the "closest and most intimate" relation between humanity and its BFF, animality.

Because they function metaphorically in Agamben, forms of zoe become largely interchangeable—humans, animals, or plants can play the role because it's a position (the abjected other) doled out by a relation to primary sovereign power, and in the end sovereign power is the real "star" of Agamben's bios/zoe drama. The differences among plants, animals, and humans as specific forms of biopolitical life are of less importance in Agamben than the fact that each is or can be a victim, a form of life that can be reduced, ignored, or killed by a dominant sovereign mode of power (wielded solely by humans). These breathless and highly elastic claims to life-as-victimization make Agamben the legitimate successor and house theorist of the late identity politics era, where, as Wendy Brown has definitively shown in States of Injury, victimization remains the primary modality of identity and recognition. And as Antonio Negri argues, in this sense Agamben's primary appeal is ideological; he everywhere reinforces that sense that sociopolitical power *actually* is what its boosters *claim* it is: top-down, to be feared, eradicating virtually all resistance and controlling nearly all life. As Negri writes, "to conceive of the relation between power and life in such a way actually ends up bolstering and reinforcing ideology. Agamben, in effect, is saying that such is the nature of power: in the final instance, power reduces each and every human being to such a state of powerlessness."[42]

And if Agamben's appeal is largely ideological (it reinforces a sense of powerlessness before sovereign power), this helps to explain why he needs to think life or bare life primarily through the constituting lens of a sovereign relation to animality: to put it baldly, consistently demonstrating a sovereign power over the life or death of plants seems substantially less dramatic, and threatens the portentously tragic tone of Agamben's entire enterprise. Remember that Yahweh, Sovereign of Sovereigns, was unimpressed with Cain's fruits of the land, requiring instead the sacrifice of his brother Abel's animal stock. And we can all recall Cain's ingenious response: if sovereignty demands sacrifice of life, guilt, and retribution, so it shall have them. In short, the animality of wild life (as zoe) remains key to the pathos and ideology of Agamben's thinking, as the proper name both for the wild space of freedom and the iron cage of power—or, more precisely, the ways in which the space of animality (the freedom deep within us) can or will inexorably morph into a concentration camp of slaughtered innocents. The bios/zoe drama is the drama of civilization and its discontents: death as edifying, sacred sacrifice (the "good" or "noble" version of our humanity as animality) versus death as mere cessation of life ("bad" or "ignoble" animality), but either way it's death nonetheless.

Likewise, Agamben's flirtation with and rejection of vegetable life as a template for zoe suggests yet another way in which the seeming centrality of the human/animal opposition harbors a whole series of hidden costs: here the primary difficulty is the way that Agamben figures the political power of bios (metaphorically the human) as purely abjecting the wild life of zoe (metaphorically, the animal). In the end this "animal" way of thinking power and life reinforces, rather than questioning, a whole series of political, theoretical, and cultural formations that are the basis for the nineteenth-century birth of biopower in Europe. The whole category of agency-as-choice, for example, is absolutely key to Agamben (the sovereign decision), as well as to animal studies: animals are ethically compelling primarily insofar as we can choose to treat them better, not to eat or slaughter them, offer them rights or protections, etc. But if plants become recognized as an ethically compelling figure for life, the whole question of sovereign human agency gets complicated substantially. In short, what's left for us to *choose* if we *decide* no longer to kill plants in order for humans to survive? To paraphrase Nietzsche, what water is there for us to clean ourselves? What festivals of atonement, what sacred games shall we have to invent, if the salad bar can no longer function as an ethical refuge from the rest of the menu at the steakhouse?

Also problematic is the fact that animal-human analogies to individual consciousness, suffering, vision, desire, and communication tend to focus all discussion on individual living beings as the only biopolitically compelling entities, the only life forms worth the name. Life that is housed in an individual being, with a rough approximation to the organism of the human (most obviously, the animal), has become the primary marker for the "other" in our biopolitical era. However, it seems more likely that individuated beings like animals are not in fact the human's most distant other but rather the closest approximation (and in fact, as Foucault shows, the subtending scaffolding) of biopower's *Homo economicus*. As we'll see in the following chapters on Derrida and Heidegger, the theoretical debate over animal lives and worlds remains a keen one. But the question of plants, whether they have lives or worlds at all, presents a series of even more interesting and difficult questions.

2

THINKING PLANTS
WITH ARISTOTLE AND HEIDEGGER

A WEALTH OF WORK IN ANIMAL STUDIES has shown us definitively that, within the canon of Western thinking about life, nonhuman animals have had a rough go of it. When animal life is discussed at all within canonical Western thought, animals are consistently downplayed, abjected, othered; animals are in fact one of the privileged names for "the other" throughout our history. As Derrida puts it, readers of Western philosophy will immediately recognize this "limit that we have had a stomachful of, the limit between Man with a capital M and Animal with a capital A."[1] One might quickly add, however, that within this twenty-five-hundred-year trajectory of thinking about living things, animals come out looking a whole lot better than plants. And, as with animals, it's not merely that plant or vegetable life is simply forgotten or ignored by a Western philosophical tradition. The cornerstones of that tradition, Plato and Aristotle, were of course very keen on categorizing all life-forms—which is to say, slotting them into their proper hierarchical places.

As with so many other Western concepts, thinking about plant life finds its initial historical contours in Plato, who writes in *Timaeus*, his book of cosmology and the creation story for Earth:

> And now that all the parts and members of the mortal animal had come to-
> gether, since its life of necessity consisted of fire and breath, and it therefore
> wasted away by dissolution and depletion, the gods contrived the follow-
> ing remedy: They mingled a nature akin to that of man with other forms and

perceptions, and thus created another kind of animal. These are the trees and plants and seeds which have been improved by cultivation and are now domesticated among us; anciently there were only the wild kinds, which are older than the cultivated. For everything that partakes of life may be truly called a living being, and the animal of which we are now speaking [i.e., the plant] partakes of the third kind of soul, which is said to be seated between the midriff and the navel, having no part in opinion or reason or mind, but only in feelings of pleasure and pain and the desires which accompany them. For this nature is always in a passive state, revolving in and about itself, repelling the motion from without and using its own, and accordingly is not endowed by nature with the power of observing or reflecting on its own concerns. Wherefore it lives and does not differ from a living being, but is fixed and rooted in the same spot, having no power of self-motion. . . . The superior powers had created all these natures to be food for us.[2]

A fair amount of what we still know today, or what we think we know today, about plant life is laid out for us here in Plato. Certainly for Plato plants are alive—the plant "partakes of life." Plants in fact have a soul (*psukhe*, often transliterated as *psyche*), but they are the "lowest" form of the living: "passive," lacking any kind of communication, awareness or sensation ("not endowed by nature with the power of observing or reflecting on its own concerns"), and sessile ("fixed and rooted in the same spot, having no power of self-motion"). All this adds up to plants' well-established role in the West as the poorer cousins to animals, the lowest threshold of living things: stuff that lives solely to serve the other, "higher" beings. As Plato starkly puts it, plants exist "to be food for us."

There is much for us to chew on within this Platonic portrait of plants as passive, sessile, lacking any motion, communication, or awareness, thereby in their essence lower than animals in the Greek tripartite scheme of living beings—the infamous "third kind of soul" within the human-animal-vegetable hierarchy of things that are alive. Research into "plant intelligence" over the past few decades has complicated or flat-out refuted almost all of this traditional picture of plant life (plants do in fact communicate with other plants; they evidence both defensive and aggressive behavior; they feign certain states to fool predators or attract pollinators; and of course plants do move, only at a much slower time-scale than most animals; there is even research to suggest that plants feel pain, or at least respond decisively to extreme danger).[3] And as I mentioned in my Preface, several very good books in an emergent "critical plant studies" have exhaustively documented the neglect of vegetable life in Western thinking, using contemporary botanical research to question this age-old portrait of vegetable life.[4] Insofar as that project of documenting what Marder rightly calls the "ethical neglect"[5] of plant life within philosophical discourse is well under way elsewhere, it is thereby not the primary task I set myself here—which is more narrowly focused on understanding how plant life functions in four of the "big" theorists whose work has circulated crucially through scholarship on animal life and biopower: Foucault, Derrida, and the duo of Deleuze and Guattari.

Some background work on vegetable life in the West will have to be done here, though, specifically as a preface to Derrida's work on plants and animals in the canon of Western philosophy. In the interest of an introductory focus (of getting to the Derrida as quickly as possible), I will emphasize here Aristotle and Heidegger on the question of life among plants, animals, and humans. I focus on them because they are the primary interlocutors in Derrida's examination of the question of animality in the text of Western thinking.[6]

As for the woeful plant, I would like to highlight one characteristic of the Western philosophical portrait of vegetable *psukhe* that I think dominates all the others. I'd like to begin in other words with what everyone within the Western philosophical canon seems to agree that plants *can* do, rather than emphasizing the laundry list of things that plants supposedly *can't* do. Putting aside for the moment the vegetable world's supposed lacks—of movement, feeling, reflexivity, and awareness—the primary positive function or essence of the vegetable *psukhe* within Greek thinking is growth itself: blind, purposeless growth, maybe, but growth nonetheless. It's this wild, potentially even cancerous, growth that is both the promise and danger of vegetable life in Greek thought. In the above quotation from *Timaeus*, for example, note how (and this is consistent throughout Plato) "wild" plants are clearly distinguished from domesticated ones—"the trees and plants and seeds which have been improved by cultivation and are now domesticated among us; anciently there were only the wild kinds, which are older than the cultivated." It is this "wild" essence of uncontrolled growth that earns plants their position as the "lowest" or most basic form of life (precisely insofar as growth and the ability to reproduce and die are the rudimentary markers for life in Western thinking).[7] As Gilbert Simondon notes in his *Two Lessons on Man and Animal*, in Greek thought "the vegetal is finalized toward generation, toward production. . . . Its growth is a growth with a view towards generation."[8]

This is perhaps clearest in Aristotle, who spends considerable time sizing up plant life and its relations to animal and human life; as Simondon wryly notes, unlike Plato, "Aristotle did not scorn the consideration of vegetal existence."[9] For Aristotle, who canonizes the notion of the "tripartite soul" (vegetable-animal-human) in *De anima* (*Peri psukhes* in Greek), plants are a problematic type of life because in the wild they have no proper end-form (no ideal entelechy). Following the classic Aristotelian scientific method, an observation of plants' being in the world reveals their nature: because they have no higher or more noble end-form (they just grow until they can grow no more), plants thereby evidence a weaker nature; they'll just feed and get bigger, toward no more ideal end, without a higher purpose. The *psukhe* of growth is, in short, all that plants have at their disposal; but, unfortunately, that nearly boundless growth-energy is too powerful for these rudimentary organisms to control or canalize. Plants simply can't control the pure immanence of growth that is their *psukhe*. As Aristotle puts it in *De anima*:

> If the movement set up by an object is too strong for the organ, the form which is its sensory power is disturbed; it is precisely as concord and tone are destroyed by too violently twanging the strings of a lyre. This explains also why plants cannot perceive, in spite of their having a portion of soul in them and being affected by tangible objects themselves; for their temperature can be lowered or raised. The explanation is that they have no mean, so no principle in them capable of taking on the forms of sensible objects but are affected together with their matter.[10]

The power of growth that is the soul of plants turns out to be "too strong" (too powerful and singular) for plants to enjoy any harmonized movement toward an entelechy or ideal end-form. Their soul, because it lacks a "mean" that might regulate its growth-power and lead it inexorably toward a higher form, brings about instead a state of dissonance within vegetable life, "precisely as concord and tone are destroyed by too violently twanging the strings of a lyre." Contrary to the contemporary view of nature (especially vegetable life) as harmonious and idyllic, for Aristotle and Plato the *psukhe* of vegetable life leaves plants in a state of constant dissonance, suffering uncontrolled growth without end. Without a higher transcendental end-form plants are merely living, growing matter.

Hence is born the familiar hierarchy of Western life, rigidly demarcated from the bottom up: plants (Aristotle's "nutritive soul") constitute mere growth itself, the lowest limit of the living, the abilities for growth and reproduction shared by all things that can be said to be "alive"; then animals, who in addition to growth (life, reproduction, and death) exhibit some higher functions or ends, such as sensation, movement, awareness of their surroundings, and appetite; and finally, of course, there's man, wielder of logos and the obvious star of the drama that is life in Western thinking. Aristotle rehearses these distinctions among forms of life in *Parts of Animals*:

> Plants, again, inasmuch as they are without locomotion, present no great variety in their heterogeneous parts. For, where the functions are but few, few also are the organs required to effect them. The configuration of plants is a matter then for separate consideration. Animals, however, that not only live but feel, present a greater multiformity of parts, and this diversity is greater in some animals than in others, being most varied in those to whose share has fallen not mere life but life of high degree. Now such an animal is man. For of all living beings with which we are acquainted man alone partakes of the divine, or at any rate partakes of it in a fuller measure than the rest.[11]

For Aristotle animals outflank plants as higher forms of life because they possess a series of obvious analogies to the highest form of earthly life, the human condition. Thereby of course the man/animal limit presents some dangers to configuring the dominance and uniqueness of the "rational animal" that is man; but these similarities also present a bounty of minute distinctions to draw, especially around the quality most prized within Greek (indeed, Western) thought: logos—reasoning and language.

From the very beginning this privileging of logos puts plants in a liminal position for Western thinking, one that's maybe even more dangerous than animals, insofar as the essence of plant life is uncontrolled growth, without any kind of obvious analogy to the higher functions of the human—to mind, reason, appetite, and so on. As Foucault has shown convincingly in *History of Madness*, animality has been a key concern in the history of reason at least in part because the animal has offered a consistently punitive portrait of what "man" would look like without the majesty of reason—a "natural" state of violence where life would inexorably be, as Hobbes famously puts it, "solitary, poor, nasty, brutish and short." So, following Hobbes, modern political theory will cue the state and a whole series of state-based interventions (prisons, laws, health and population measures) to mitigate this "problem" of man's animal nature. As we saw via Foucault in Chapter 1, animals are figured as other not because of their absolute distance from modern definitions of "man" but precisely because of their intimate, hidden subtending proximity. It's plants that are the odd other out, included in the pantheon of life, but bereft of life's higher functions, thereby unable to control their own violent and intense growth, much less mitigate it in or for others, so plants likewise have no political existence.

The Power of Soul

Aristotle's *Peri psukhes* suffers immensely in its Latin title, *De anima*, precisely because *soul* (translating the Latin *anima*, which in turn is translating the Greek *psukhe*) carries with it unfortunate later baggage from Christianity: the soul as the little light deep within you, the real you, the trace of God. "Soul," in short, gets bound up with questions of the afterlife and, even more so, individual identity; it seems something you possess (maybe even the only thing you possess in the end—so much so that you can sell it to the devil, if need be). In Christian parlance the soul becomes the essential you that remains when everything else (including the materialities of the body or the earth) is stripped away. The Greek *psukhe*, however, carries few of those Christian valences and little to none of that kind of individualist sense of identity. The Greek "essence" that is *psukhe* is probably better understood as a vital principle or, better yet, the capacity for a certain kind of "life."[12]

In book 2, after an exhaustive review of other thinkers' positions on the soul in book 1, Aristotle gets down to business, asking "What is soul [*psukhe*]?" (*DA* 412b10). Like the later Christian soul, Aristotle suggests that the *psukhe* is what makes a thing what it is, a kind of essence, but that essence is more of an exercisable power or potential than it is an interior thing or possession. Aristotle uses the example of the eye and its "soul": "when seeing is removed the eye is no longer an eye, except in name" (*DA* 412b20): seeing is the *psukhe* of the eye—its ownmost potential. When a thing's life-principle is removed (as with a human corpse, for example), it is no longer that thing, except in name. In short, "as the pupil plus the power of sight constitutes the eye, so the soul plus the body constitutes the animal" (*DA* 413b2)—with the *psukhe* here glossed or exemplified as "the power of sight" (the ability of the eye to do something in particular, to see) rather than understanding the "soul" of seeing as a largely static or timeless essence of sight to be found deep within the eye.

We see perhaps a helpful hint concerning Aristotle's *psukhe* in an English translation issue surrounding his discussion of the "nutritive soul" of plants: Aristotle writes, "This power of self-nutrition can be separated from other powers mentioned, but not they from it—in mortal beings at least. The fact is obvious in plants; for it is in fact the only psychic power they possess" (*DA* 413a32–34). The contemporary reader of English perhaps stumbles here over the meaning of plants' "psychic power" (they can tell fortunes?), but it is a literal and faithful translation of the adjectival form of *psukhe*: nutritive growth is the only "power of life" that plants have at their disposal. While other forms of life can enact additional "higher" powers (sensation and movement in animals, thinking in humans), no life-forms can exist without the nutritive *psukhe* of plants, and concomitantly all living things necessarily deploy a dose of the nutritive or vegetable soul, what Aristotle calls "the originative power, the possession of which leads us to speak of things as *living* at all" (*DA* 413b1): "The nutritive soul is found along with all the others and is the most primitive and widely distributed power of soul" (*DA* 414b24).

When Aristotle asks, "What is the soul of plant, man, beast?" (*DA* 414b33), he means to ask not so much about the otherworldly "essence" of these differing life-forms but about the differing capacities and abilities of vegetable, animal, and human life. In other words, when Aristotle asks, "What is its *soul*?" he primarily asks after its *abilities*. In this way a thing's "ends" or entelechy (what it does) reveals its "essence" (what it is) (*DA* 415b18–21). So when one of Aristotle's students reports an observation that I cited earlier from *On Plants*, that "a plant which is fixed in the ground does not like to be separated from it," he's not so much taking a moral stand or anthropomorphizing plants by ascribing feelings to them as he is merely following out Aristotelian method: observing that when you pull up a plant, that interrupts its *psukhe*, its primary capacity for life, its nutritive functioning.

This Aristotelian insistence that the soul is what it does (or that essence is only recognizable through an examination of ends) has functioned historically as both the glory and curse of the vegetative soul: all living things have some measure of the vegetative soul, the reasoning goes, or else they would not live or grow at all. You can see that something partakes of the vegetative soul precisely insofar as it lives—it emerges, grows, has the potential to reproduce, and passes away. But because plants have no observable end, no way to channel that growth toward higher ends, their *psukhe* is not "noble." Plants are certainly alive in Western thinking; no one disputes that. Indeed, the vegetative soul has for centuries constituted the threshold limit of everything living. Every schoolboy for a thousand years knew from Aristotle that we humans possess a tripartite soul: the vegetative soul of growth, the animal soul of sensation, and the human soul of reason (which routes the other two toward higher powers).

In our era of biopower we routinely summon Aristotle's twenty-four-hundred-year-old definition of man as the rational or political animal—an animal with an additional capacity for rational thought and political existence. The Aristotelian role of plants or the vegetable soul within our makeup remains, however, somewhat less commented upon. As Aristotle writes in *Parts of Animals*:

> For plants get their food from the earth by means of their roots; and this food is already elaborated when taken in, which is the reason why plants produce no excrement, the earth and its heat serving them in the stead of a stomach. But animals, with scarcely an exception, and conspicuously all such as are capable of locomotion, are provided with a stomachal sac, which is as it were an internal substitute for the earth. They must therefore have some instrument which shall correspond to the roots of plants, with which they may absorb their food from this sac, so that the proper end of the successive stages of concoction may at last be attained. The mouth then, its duty done, passes over the food to the stomach, and there must necessarily be something to receive it in turn from this. This something is furnished by the bloodvessels [*sic*], which run throughout the whole extent of the mesentery from its lowest part right up to the stomach. (*PA* 650a20–26)

For Aristotle the stomach constitutes both animal and human life's analogue to the function of the "ground" for plants—we carry "an internal substitute for the earth" within us, located primarily in the "stomachal sac." Our locomotion is made possible by such a portable earth, which is to say a vegetable soul or power of growth that moves along with us. Likewise, Aristotle is very clear that our circulatory systems are plantlike—they "correspond to the roots of plants" insofar as they spread the *psukhe* of growth and nutrition throughout the animal organism. In short, for Aristotle we are not only rational animals; we are also walking plants.

In any case Aristotle's *De anima* codifies much of what Western thinking has to say about "life," and that script is primarily acted out in a drama of threes: (1) The nutritive or vegetative form of life: the most basic property is shared by all living things but is insufficient in itself as it is sessile; it doesn't move but merely channels the principle of (uncontrolled) growth. Plants are, one might say, pure immanence, bereft of transcendence. (2) Animal life: this form has some measure of movement and sensation, and has many other analogues to the human: the rudiments of negation, control, or cunning usage of natural instincts (the wolf doesn't always attack, as the plant supposedly always grows; the wolf stalks and picks its spots). Animals, then, remain in a liminal position concerning life and the human, with some "higher" powers and plenty of desiring appetite but no reason to speak of. And finally (3) human life: this form has access to logos (language and reason), thereby has an opening to higher concepts (communication, negotiation, science, self-knowledge) and likewise to a reason-based political existence (as opposed to the merely instinctual pack or herd mentality of animals).

This tripartite scheme remained intact in the Middle Ages, especially in Aquinas, through the early modern period (Shakespeare references it several times), and in fact right up to Heidegger—who reinscribes Aristotle within the Heideggerian tripartite scheme of life, where the stone is "without world," while the animal is "poor in world" and the human "world-producing." A lot of ink has recently been spilled within animal studies on this Heideggerian world-picture (a good bit of it by Derrida), so it's Heidegger's text that I think requires the most torque by way of introduction or background to the question of life distributed among the animal, the earth, the plant, and the human. But I take this detour through Aristotle (clearly, Heidegger's source-text in his thinking about "life"—if not the source for his thinking about virtually everything else)[13] so we can note from the outset the strange and nearly complete disappearance of vegetable life from contemporary work on biopower (even though direct discussion of plant life is all over Aristotle). It's not so much that plants have been excluded from the tradition of Western thinking. But in Heideggerian parlance it seems that the philosophical question of plant life has today been largely forgotten.

Plant Characters:
Plant Life in Heidegger's 1929–1930 Lectures

If Aristotle's *De anima* is the most famous baseline ancient Western text on the topic of life, Martin Heidegger's 1929–30 lecture course *The Fundamental Concepts of Metaphysics: World, Finitude, Solitude* constitutes a founding text for much contemporary philosophical work on biopower and animal studies, and it functions as a nearly obsessive site for Derrida's cornerstone work on animality. Derrida begins thinking about this Heidegger lecture course shortly after it's published in 1983; he first takes it up in 1987's *Of Spirit* and comes back to it time and again in his voluminous work on animality.[14] Why all the obsessive focus on this one text? In this lecture course Heidegger is interested in the most basic questions about life, the "Fundamental Concepts," or what he calls "the living character of the living being": "In what way should life, the animality of the animal, and the plant-character of the plant, be made accessible to us? . . . How are living beings as such—the animality of the animal and the plant-character of the plant—originarily accessible?"[15] As is clear here, Heidegger wants to approach this question of life primarily through the question of world, which he defines quite simply "as the accessibility of beings" (*WFS* 268), thereby staying close to Aristotle's refinement of the *psukhe*, which accrues to this or that form of life. In short, the question of life is this: to what powers does X or Y life-form have access?[16]

So, for example, early on Heidegger asks after the question of access to sleep in terms of differing life-forms: "We do not say that the stone is asleep or awake. Yet what about the plant? Here already we are uncertain. It is highly questionable whether the plant sleeps, precisely because it is questionable whether it is awake. Yet we know that the animal sleeps. . . . Yet the question remains as to whether its sleep is the same as that of man" (*WFS* 62).[17] Heidegger even offers a kind of commonsense definition of life, one that he will both work with and against: "life—that is, the kind of being that pertains to animals and plants" (*WFS* 191). In the end, though, it's "world" and not "life" that becomes the linchpin question for Heidegger. As he contends, "the main thrust of our considerations does not rest upon a thematic metaphysics of life (of plants and animals)" (*WFS* 193) but on the question of world—that is the refashioned question of *psukhe* or capacity for life as a question of world or "access."

Heidegger's *World, Finitude, Solitude* has become a kind of fetish text for contemporary biopolitical work, especially in its most basic premises: "the main points of our approach are encapsulated in three theses: [1.] The stone is worldless; [2.] The animal is poor in world; [3.] Man is world-forming" (*WFS* 184). If world is "the accessibility of beings," then one might initially translate: stone has no access to things or beings; the animal has some constricted access to things or beings; while the human, one might say in a Heideggerian turn of phrase, has access to the question of access itself—to what he will call elsewhere the "worlding of the world." All this is very interesting and controversial, and could keep us focused here for many, many pages, untangling and contesting Heidegger's hierarchy among the various life-potentials that supposedly accrue to stones, animals, and humans. We could also discuss for quite some time what exactly Heidegger is doing with Aristotle here, as Heidegger clearly states his debt: "Aristotle grasps the concept of life in a very broad sense that includes the being of plant, animal and human. Book Three of this treatise [*De anima*] deals precisely with living beings in the sense of man. The distinguishing feature of man is the logos" (*WFS* 313), and Heidegger goes on to state quite straightforwardly that "what we call here world-formation is also the ground of the very inner possibility of the logos" (*WFS* 335). Indeed, recent work in biopower and animal studies looks closely at Heidegger's 1929–30 lecture course precisely because of the controversial comparative hierarchy of "life" that it proposes: Heidegger's canny updating and reinscription of Aristotle's tripartite soul—life understood in terms of futurity and access to possibility, not in terms of a biological substrate or essentialist set of distinctions between the living and the nonliving.

I want to focus, however, on a simpler question in Heidegger's analysis: what happened to plants in all this? Heidegger is explicitly working through Aristotle, who as we have seen has much to say about life and the plant *psukhe* that distinguishes living from nonliving things. And Heidegger himself will, as we have seen, ask after the "plant-character of plant life" or will ask whether plants sleep or not. The plant consistently comes up in Heidegger's discussions of life-forms, only to be almost immediately elided. For example, when introducing the method of his "comparative analysis," Heidegger begins:

> Man has world. But then what about the other beings which, like man, are also part of the world: animals and plants, the material things like the stone, for example? Are they merely parts of the world, as distinct from man who in addition has world? Or does the animal [who we note here has lost her friend the plant since the earlier sentence] too have world, and if so, in what way? And what about the stone? However crudely, certain distinctions immediately manifest themselves here. We can formulate these distinctions in the following three theses: [1.] the stone (material object) is worldless; [2.] the animal is poor in world; [3.] man is world-forming" (*WFS* 177).

One might add, or one might wonder why Heidegger doesn't add: [1.5] About the plant, we're not so sure.

Clearly, in Aristotle, as we have seen, the plant soul is the limit of life itself—the nutritive *psukhe* is the one thing that all living things can and must perform in order to survive. Plants are for Heidegger likewise clearly alive, "as distinct from the non-living being which does not even have the possibility of dying. A stone cannot be dead because it is never alive" (*WFS* 179). So is the plant merely subsumed into the animal for Heidegger, both as forms of life that are nevertheless poor in world? Insofar as "the animal's way of being, which we call 'life,' is not without access to what is around it and about it, to that amongst which it appears as a living being" (WFS 198), the animal in Heidegger certainly does retain some restricted (poor) access to relationality and so to world. But what about the plant's way of being, which we also call life? The intention of the lecture course is "to free us completely from the naive view from which we originally started, namely that the beings in question—stone, man, animal, and indeed plants—are all given on the same level in exactly the same way" (WFS 207), and we have ample discussion of three of these four forms, in terms of their access to world. One wonders, do plants have access to what is around them? Surely it would seem less so than animals (though Heidegger's analysis of the bee, which doesn't act but merely makes instinctual movements, might suggest otherwise); but just as surely plants have access to world (the possibility of change, growth, withering) more so than the stone—which never grows or dies, Heidegger contends, because it never feeds or lives. Plants live and die, turn toward the sun, grow, and reproduce; but what is the character of that "world"? On this point, Heidegger remains resolutely silent.[18]

In fact, in a very odd twist, while the plant in Heidegger's 1929–30 text remains unthematized, he does offer a discussion of another liminal force of life, somehow stuck between the material object of the stone and the mobile (but world-poor) animal: namely, the status of things, or "equipment" [Zeug] in Heidegger's vocabulary. Somehow, it's equipment, not the plant, that finds itself positioned between the stone and the animal. Heidegger writes,

> As soon as we attempt to clarify the essence of the organism, we immediately find ourselves confronting a whole range of different kinds of beings: purely material things, equipment, instrument, apparatus, device, machine, organ, organism, animality—how are these to be distinguished from one another?
>
> Is this not the same question we have been striving to elucidate all along, except that we can now insert beings having another specific manner of being—namely, equipment, instruments, machines—between the piece of material substance (the stone) and the animal? (WFS 213).

In Heidegger, then, it's not the vegetal mode of being that is somehow situated between world-poor animal and worldless stone, but the machinic mode of being that takes up its home in that interval. How so? He goes on to argue, based on his famous analysis of the hammer in section 15 of Being and Time, that pieces of equipment "are neither simply worldless like the stone, nor are they ever poor in world [like the animal]. Yet presumably we must say that equipment, articles of use in the broadest sense, are worldless, yet as worldless belong to world. In general this means that all equipment—vehicles, instruments, and especially machines—is what it is and in that way that it is only insofar as it is a product of human activity. And this implies that such production of equipment is only possible on the basis of what we have called world-formation" (*WFS* 213). So equipment finds its place within world "insofar as it is a product of human activity," which is to say that if the human is indeed characterized by its abilities for world-building, it's going to need some tools to do the building.

This in fact is the ground cleared for the hammer in *Being and Time*—which Heidegger refers us to in the 1929–30 lecture course, when he abruptly breaks off consideration of equipment. In Heidegger's world one does not go from the hand to the hammer (from presence to presence, the so-called "vulgar" concept of time), but there is always the intervening step of the project: the thrown, future possibility that the hammer is being sought out in order to fulfill. Dasein's primary relationship is not with tools or things (nor even, as Emmanuel Levinas consistently points out, with another Dasein) but with possibility—which is to say a primary and unique relation with the world, ultimately with Being itself. And Dasein seeks out tools to help her fulfill that projected future potential of world: to build shelter, libraries, concepts. In *Being and Time*'s vocabulary, the world is the prior, dominant, or *zuhanden* (ready to hand, potential) field in which the *Vorhandenheit* (presence to hand) of tool usage comes into play. And this analysis of world and tool moves hand in glove with Heidegger's critique of temporality: any state of presence is derived from a prior state of possibility (*techne* from *poesis*, *vorhanden* from *zuhanden*, tools from world, Dasein from Being). So objects like the hammer (more than the plant?) remain in relation to world but only insofar as the hammer or other tool finds itself constantly swept up in the future- and possibility-oriented movements of Dasein.

In short, as Graham Harman argues in *Tool Being: Heidegger and the Metaphysics of Objects*,[19] it may turn out that being-with-tools or objects is the most intense mode of Heideggerian being-in-the-world, precisely because the hammer (or more accurately, the broken hammer) is the shortest and most privileged route toward a glimpse of Being for Dasein: the broken hammer reveals being-in-the-world as characterized by multiple possibilities (in short, it reveals that there are many ways to drive a nail) rather than the singular end-orientation of presence to hand. Oddly, solitary work with tools seems to retain more of the sense of robust ontological possibility for Dasein than does engagement with other people, the infamously inauthentic idle chatter of *das Man* in *Being and Time*. So Dasein's work with tools functions somewhat similarly to its work with (poetic) language, as a royal road to the disclosure of worldly possibility—and the world as possibility itself. In short, it is through action in the world and interaction with possibility that Being is revealed to Dasein. As my teacher Tom Sheehan used to say, in Heidegger "excess gives access."[20]

While tool-being has a privileged role in this revelation process for Dasein, animals (and even more so plants) don't. Because animals are not (merely) tools—they are alive—they can't unproblematically be annexed as props within Dasein's world. Animals have a limited world of their own, insofar as animals display "drives" and "capacity" (*WFS* 228), but not "action" (*WFS* 236): "the behavior of the animal is not a doing and acting, as in human comportment, but a driven performing [*Treiben*]. In saying this we mean to suggest that an instinctual drivenness, as it were, characterizes all such animal performance" (*WFS* 237). Heidegger insists that the animal is captivated by its environment, rather than forging relations (acting or exercising agency) within its world. Heidegger dubs captivation and benumbment [*Benommenheit*, benumbment or bedazzlement] as "the inner possibility of animal being itself" (*WFS* 239) and insists in the end that "the animal behaves within an environment, but never within a world" (*WFS* 239).[21]

In his most extended example of animal behavior in *World, Finitude, Solitude*, Heidegger takes up an analysis of the bee, following experiments performed by Jakob von Uexküll in his founding biosemiotic work on the *Umwelt* (environment, literally "around-world") of animal perception. Cashing out his claim that animals behave rather than act (they're "in" a world but don't have access to the means of transforming it), Heidegger argues that the bee is captivated by its surroundings, and here he refers explicitly to Uexküll, who showed that if you cut a feeding bee in half, it continues to suck until the food source is gone (*WFS* 242). Heidegger uses the bee analysis to show that the animal goes from one instinctual drive to another: the bee does not decide, or act, or transform its world, and likewise the bee is not easily distracted from the tasks it's driven to perform—even vivisection goes unnoticed. The bee is merely driven to perform certain instinctual tasks within its environment: as the bee finds food, takes it in, and returns to its hive, we see "the drivenness of the capability is redirected into another drive" (*WFS* 243) time after time.

I'm certainly not a biologist or an entomologist, and here I am not going to argue against this picture of animal life that behaves rather than acts. But I would merely point out that, if we accept this Heideggerian analysis of animal life, then it seems that plant life (or what's traditionally been said in the West about plant life) lines up almost term for term with the characterization of animals: plants, like animals, have instinctual drives but not worldly responses; like animals, plants are held captive by their environments—light, water, food, territory, and so on—but exercise no definitive transformative agency over that *Umwelt*. In Heidegger it seems that the plant is right alongside the animal in sharing a world-poor status.

Following Uexküll, Heidegger extends his example of animal captivation by recounting an experiment in which researchers moved a hive after bees had gone foraging out of it. Inevitably, the bees fly back to where the hive originally was located (not to its new location). The lost bees then look around, and eventually discover the changed hive placement. Heidegger comments on this experiment and its relation to world: "bees can find their way back over great barren distances with no . . . orienting features. What is it that guides them and keeps them, so to speak, in the right direction? Neither the color nor the smell of the hive, neither landmarks nor other such objects by which they could take their bearings, but—what, then? The sun" (*WFS* 245). So animals, like plants, finally only have a primary captivating and guiding relation with one entity: the sun.[22] Heidegger continues: "The bee is simply given over to the sun and to the period of its flight without being able to grasp either of these as such, without being able to reflect upon them as something thus grasped. The bee can be given over to things in this way because it is driven by the fundamental drive of foraging. It is precisely because of this drivenness, and not on account of any recognition or reflection, that the bee can be captivated by what the sun occasions in its behavior" (*WFS* 247). Here, as above, I'm less interested in arguing with Heidegger's picture of animal life than I am in pointing out that if you are going to argue with Heidegger about the *weltarm* animal captivated by and benumbed within its surroundings (or, following Derrida, if you wonder whether the human can be said to "possess" any of the things denied to the animal), then it seems you're likewise committed to arguing with Heidegger about the status of the plant as well, insofar as Heideggerian poorness in world comes down to a life-form's being captivated by (and prisoner to) its environment and instincts: living, behaving, ceasing to live, but having no access to the "as-such" that might understand or transform that life (primarily through logos and an awareness of death) into something "more." Neither the animal nor the plant has much in the way of access to excess.

In the end Heideggerian animal life, in its world-poorness, remains largely identical to plant life insofar as both are rooted in and captive to their respective environments. Indeed, Heidegger comes to the conclusion that the environment is a ring that encircles and benumbs the animal, not a world thru which it moves: "the animal's behavior in relation to the sun does not occur as a form of recognition which is subsequently followed by an appropriate form of action. Rather, the animal's captivation by the sun only occurs in and through its instinctual foraging drive" (*WFS* 249). The animal *psukhe*—to forage, to be captivated by its environment, to follow its instincts—is what turns it toward the sun, toward the food sources, toward its limited but ownmost possibilities of survival. However, aside from inhabiting a wider range of benumbing encirclement (animals are driven and able to do their instinctual "foraging" over much wider territories than plants), there is nothing said here about animal life that wouldn't also accrue to plants: oriented first and foremost by the sun, animals as well as plants forage their environment for sustenance—plants' roots extend toward water, the leaves turn toward light, the blossoms open to the bees. The plant of course is enthralled to its immediate environment even more so than the bee, and the plant doesn't display many obvious analogues to human consciousness (as many "charismatic" animals could be said to model); but these questions of analogy are not theoretical crux points for Heidegger (any more than sleep or a wide range of movement is).

Finally, Heideggerian world-poorness characterizes a life-form that is alive but remains captive to its surroundings, without any access to anything "as such" (not to life or, even more especially, death): for the animal "there is no apprehending, but only a behaving. . . . The animal is precisely taken by things" (*WFS* 247). Given this sense of animal life, Heidegger's strange elision of the plant may be easier to grasp, then: if the distinction between Dasein and other forms of life will be one of access to world (as such) versus mere enthrallment to an environment, then plant life would constitute too "easy" a target: it requires a series of arguments to demonstrate that the animal is enthralled to its surroundings, while the plant seems a prisoner of its environment on the face of it—lacking thought, mobility, and behavior. Insofar as there are precious few (if any) analogues to humans' "world-forming" powers, perhaps the plant doesn't pose for Heidegger the same set of argumentative possibilities as the animal.

In any case, as we will see, it's just this Heideggerian argument concerning animality and world-poorness that Derrida will latch onto in his deconstruction of Heidegger: the distinction between man and animal remains for Heidegger based on access to the "as-such" (to reflexivity, knowing, dying rather than ceasing to exist). Derrida will dispute this premise concerning the as such less than he will dispute the dubious sense that humans' movements (our drives, actions, or instincts) are any less "enthralled" to their environments than animals' movements.

However, here's my overarching question concerning this analysis: if the animal is barely distinguishable from the plant in Heidegger (alive, but held captive by its environment), then Derrida's wondering whether humans can separate themselves from animals is also asking whether we can separate ourselves from plants. Indeed, Heidegger himself opens the door (or rolls away the stone at the mouth of Plato's cave or Jesus's tomb) when he wonders whether animals can tell the difference between a "real" light source (a flame, the sun) and a mere reflection: Heidegger notes that "Moths are light-seekers and thus seek out illuminated surfaces rather than the intensity of the source of light" (*WFS* 251)—which is at least partially to say, moths and other animals can't tell the difference between light "as-such" and false sources (say, reflected in a mirror, or on the wall of a cave, or for that matter on a 3D home movie screen). This is to say that the very situation that Heidegger dramatizes for us (the animal's being unable to tell the difference between a true light source and mere reflections) constitutes a repetition of one of the oldest philosophical parables in the West, Plato's allegory of the cave—where the unphilosophical masses can't tell the difference between the things themselves (illuminated by the real light of the sun in the higher world of knowledge) and the manipulation of mere reflected images in the dark cave of ignorance. In other words, if for Heidegger the primary problem of animal life is enthrallment (to the false light of drives rather than to the as-such of possible transformation), then how are humans in Plato's cave (that is to say, most of the species) different from animals? And if animals are the unfortunate world-poor siblings of plants, how then are humans different from plants? The plant that therefore I am?

Additionally, and perhaps more pressingly, this thematic of environmental benumbment and some life-forms' (in)ability to separate true light (the "as such") from mere captivating reflection opens up a more disturbing question about life nested within Heidegger's political thinking of the 1930s: how to tell the difference then between authentic Dasein and *das Man* (the unthinking mass, literally "the they")? In 1929–30 Heidegger is of course only three years out from what will become his most infamous public endorsement of Nazism, his Rector's Address. And one might legitimately wonder whether this course on animality constitutes a skeleton key to Heidegger's embrace (however tenuous or short-lived) of Nazism—which is of course a biopolitics of legitimate and worthy forms of life and the gulf that separates them from lower, unworthy life-forms.[23] Whatever else it was, Nazism was a system set up by and dedicated to those who were confident that they could tell the real light from the illuminated surface—the difference between idle chatterers and prisoners of technology, watching projections on the cave wall, and the authentic Dasein who has seen the real light of the open *Lichtung*. Indeed, one might ask, flat out: on Heidegger's terms is *das Man* finally little more than an animal, or maybe even a plant—poor in world, captivated by its environment, bereft of access to excess, fit not to die authentically but only to perish?

Heidegger's *Black Notebooks* offer some chilling food for thought in this way, as they confirm that his acceptance of National Socialism goes back to 1930—just as he was delivering these lectures on biology. In his *Black Notebooks* Heidegger writes "Thinking purely 'metaphysically' (that is, in terms of the history of Being), in the years 1930–1934, I took National Socialism as the possibility of a crossing-over to another beginning and gave it this meaning."[24] Gregory Fried comments on this passage: "Not only does Heidegger confirm his *philosophical* understanding of National Socialism's significance—at least until 1934—he also acknowledges how *early* he came to believe this: 1930, three years before the Nazis came to power."[25] Likewise, the *Black Notebooks* tend to suggest that Heidegger's anti-Semitism had everything to do with a conception of the German people's authentic, rooted worldhood versus the wandering, benumbed cosmopolitanism of their enemies in both world wars: Heidegger calls the Germans a "people of authentic historical force," in contrast to what he dubs the "metaphysical inanity" of the Jews, the Americans, the English, the French, and the technology-obsessed Russians. "Only the German can give new poetic voice to Being," he writes.[26] As David Krell sums up Heidegger's thoughts on "World Jewry" in the *Notebooks*, however, "Most disturbing in his account of *Weltjudentum* is the word *Weltlosigkeit*, which Heidegger used in 1929–30 to describe the lifeless stone; in these notebooks he is liable to include the animal in such deprivation of world. And now the worldless, lapidary animal is joined by a worldless world-Jew." As Krell straightforwardly puts it, for Heidegger "The Jew lacks world and earth."[27]

This being the case, the 1929–30 discussions of animals and their poorness in world begin to look like a bridge to a common and vicious form of nationalist anti-Semitism, with both animals and enemies of the Reich lacking worldhood. In other words this link to the biological in the 1929–30 lecture course (to the question of which specific forms of life have access to world and thus to life itself) helps us to think more about something that, of course, we'll never understand. Perhaps, as Roberto Esposito writes, "This—and nothing else—was Heidegger's Nazism: the attempt to turn himself directly towards the proper, to separate it from the improper, to make it speak in the affirmative, originary voice."[28]

In any case, with this discussion of Aristotle and Heidegger's 1929–30 lecture course as background, we turn now to Derrida's discussions of life, world, and animality—all of which are overtly (de)constructed on this Heideggerian scaffolding.

3

ANIMAL AND PLANT, LIFE AND WORLD
IN DERRIDA; OR,
THE PLANT AND THE SOVEREIGN

IT'S A LITTLE ODD THAT JACQUES DERRIDA has become such a linchpin
figure in the field of animal studies. Indeed, it's ironic (though not I think inaccurate) to suggest that if animal studies publications continue coming out at
their present pace, Derrida will be remembered decades from now primarily
as a thinker of animals rather than as a theorist of *écriture* or a deconstructor
of Western metaphysics. I suppose there's nothing particularly strange in the
fact that his late essays collected in *The Animal That Therefore I Am* became
a jumping-off text for Cary Wolfe's foundational academic work on animality and that Derrida remains a focal point for countless other theorists of the
animal who've followed in Wolfe's tracks.[1] The oddity, perhaps, is that Derrida
posthumously finds himself lauded in the North American *popular* press when
the discussion turns to animals—in, for example, Jonathan Safran Foer's best
seller *Eating Animals* (where Derrida is approvingly quoted a few times), and
even in a place where he never got any love when he was alive, the *New York
Times*. While making it clear that they are sticking to their script (grumbling
that "his writing is almost impossible to capture in a quotation"—referencing
their own infamous 2004 obituary headline for Derrida, "Abstruse Theorist
Dies at 74"), by 2012 the *Times* informs us, however begrudgingly, that "Jacques
Derrida has had an equally strong influence" as Peter Singer when it comes to
"the way we think about animals."[2]

Throughout *The Animal That Therefore I Am* Derrida goes to great lengths
to point out that he had been interested in the question of the animal for a long

time. Over several pages he summarizes the "horde of animals," the "innumerable critters" that run through his texts (*ATT* 37–41). Though for someone seeking a review of all the forms of life treated in Derrida's work (including, say, plant life, which we're interested in here), his own summary in *The Animal* text remains a bit strange. For example, in beginning to review his own work on "the question of the living and of the living animal" (*ATT* 34), Derrida goes first and directly to "White Mythology," an essay that I had always taken to be largely about the heliotropic—which is to say, vegetal—nature of metaphor in the text of Western thinking (metaphorical tropes as those flowers of rhetoric that turn toward the sun of truth). In "White Mythology" Derrida points out that Western thinkers since Aristotle have "been carried along by the movement which brings the sun to turn in metaphor; or have been attracted by that which turned the philosophical metaphor towards the sun. Is not this flower of rhetoric [like] a sunflower? That is—but this is not an exact synonym—analogous to the heliotrope?"[3] Derrida goes on to note that the plant life of metaphor is both inside and outside the logos that defines *anthropos* for Aristotle. Metaphoricity (the movement from the sensible to the intelligible) makes understanding possible but simultaneously makes it impossible for intelligibility fully to escape its dependence on these flowers of rhetoric, which are always already overgrown. As Derrida points out, "Each time polysemia is irreducible, when no unity of meaning is even promised to it, one is outside language. . . . At the limit of this 'meaning-nothing,' one is hardly an animal, but rather a plant" ("WM" 248).[4]

In *The Animal* book, however, Derrida suggests that "White Mythology" is not primarily about language's plant figures but rather about what he later names "the animality of writing" (*ATT* 52): "As I see it, one of the most visible metamorphoses of the figural, and precisely of the animal figure, would perhaps be found, in my case, in 'White Mythology'?" (*ATT* 35).[5] Indeed, even when Derrida recalls the founding biblical story of murderous intrahuman antagonism (Cain and Abel), and notes its originary connection to animal sacrifice, he seems oddly uninterested in the (even more abject) status of plant life within this foundational story: "Cain, the older brother, the agricultural worker, therefore the sedentary one, submits to having his offering of the fruits of the earth refused by God who prefers, as an oblation, the first-born cattle of Abel, the rancher" (*ATT* 42). While certainly the story of Cain and Abel is about the fundamental role of animal sacrifice at the basis of Judeo-Christian relations to the human other, as well as to the divine—it demonstrates, in Derrida's concise phrase, that "politics presupposes livestock" (*ATT* 96)—it's also a story about the abjection of plant life in forming that crucial relation of value and sacrifice (and thereby the plant's elision within the whole apparatus of "life-as-sacrifice" that, as Agamben has argued, is crucial to Western thinking and politics). In short, it is Yahweh herself who opens this foundational abyss between a privileged, value-laden animal life (whose sacrifice *means* something—is maybe even the condition of possibility for religious and cultural meaning) and a necessary but abjected other of sessile plant life (a form of life that is essentially meaningless). As Derrida reminds us, God "refused Cain's vegetable offering, preferring Abel's animal offering" (*ATT* 43), and thereby we see inaugurated a linchpin of Judeo-Christian culture: animal sacrifice is the only sacrifice worth the name. As Derrida notes in *Glas*, from the beginning it may be that "the flower is in the place of zero signification."[6]

In the course of his *Beast and the Sovereign* lectures Derrida wonders out loud several times about whether what he has to say about animal life might be extended to plant life as well. For example, he opens the second volume of lectures with a kind of summary of the first, and its questioning of any category of "the animal in general"; but he just as quickly insists that there is common ground among animal and human lives: "once we have given up on saying anything sensible or acceptable under the general singular concept of 'the' beast or 'the' animal, one can still assert at least that so-called human living beings and so-called animal living beings, men and beasts, have in common the fact of being living beings (whatever the word 'life,' bios or zoe, might mean, and supposing one has the right to exclude from it vegetables, plants and flowers)" (*BS* 2:10). Here Derrida clearly marks the exclusion of vegetal life from the discussion of "living beings" but leaves it unexamined. Derrida even more keenly notes several times the suspicious eliding of plant life in Heidegger's famous 1929–30 lecture series *World, Finitude, Solitude* (where, as we saw in Chapter 2, Heidegger holds that the stone is worldless, the animal poor in world, and man world-forming). At one point, wondering aloud about "the ambiguity of vegetables and plants" in Heidegger's schema, Derrida asks flat out the question I noted earlier: "Would Heidegger have said that the plant is *weltlos* like the stone, or *weltarm* like the living animal?" (*BS* 2:6).

For an answer to this excellent question we look to Derrida's next sentence: "Let's leave it here for now: the question will catch up with us later." When it does reappear later in the lecture series, Derrida continues: "Heidegger wonders more than once how life is accessible to us, be it the animality of the animal or the vegetable essence of the plant; *and twice—this is highly interesting in my view—Heidegger classifies the plant, the plant-being of plants, the vegetable, as they say, among the phenomena of life, like the animality of the animal; but he will never grant to the living being that the plant is the same attention he will grant the living being that the animal is*" (*BS* 2:113–14, my emphasis). OK, you might think, now we're onto something. But Derrida's next sentence, following the protocol already established, immediately closes this parenthetical observation and leaves the question of vegetal life completely, and oddly, in abeyance. (The next sentence reads, "So how is life accessible to us, given that the animal, Heidegger notes, cannot observe itself . . . ?")

Similarly, while discussing iterability, repetition, and originality within his commentary on Heidegger, Derrida offers a brief aside on cloning:

> *Klôn* is, moreover, in Greek, like *clonos* in Latin, a phenomenon of *physis* like that young sprout or that (primarily vegetable) growth, that partheno-genetic emergence we talked about when we were marking the fact that, before allowing itself to be opposed as nature or natural or biological life to its others, the extension of *physis* included all its others. There again it seems symptomatic that Heidegger does not speak of the plant, not directly, not actively: for it seems to me that although he mentions it of course, he does not take it as seriously, qua life, as he does animality. (*BS* 2:75–76)

Interesting. In the next sentence Derrida muses over the ranunculus, an aquatic flower whose name means little frog. Then right back to *Robinson Crusoe*. Nothing more on the exclusion of the plant in Heidegger.

Turning back to *The Animal That Therefore I Am*, the single reference to plants themselves is raised within another citation I've already highlighted, where Derrida quotes a passage I've previously cited from Heidegger's 1929–30 lecture course: "We do not say the stone is asleep or awake. Yet what about the plant? Here we are already uncertain. It is highly questionable whether the plant sleeps, precisely because it is questionable whether it is awake" (*ATT* 148). At this juncture Derrida tantalizingly comments, "one should spend a long time on this" (*ATT* 148) question of sleep and vegetable life raised by Heidegger. But rather than do so, he again abandons the question immediately, and without further comment, to stay on the trail of the animal. It seems that when his discourse focuses on questions of animal life, the Derridean stage direction for discourse surrounding plant life remains akin to Shakespeare's famous directive in *The Winter's Tale*: "*Exit, pursued by a bear*" (3.3).

At one level it's completely understandable to background the question of plant life within an inquiry dedicated to animal life (to follow a singular trail, to see where it leads specifically), but there are other places where Derrida seems not so much to wonder about but actually to *further* this exclusion of plants from the larger discussion about life. For all of his relentless interrogation of Heidegger's views on animality and its discontents, Derrida's late work follows and endorses Heidegger's insistence that the question of life is entangled at all levels with the question of world—which is to say, a singular being's relation to possibility, futurity, and to itself (as an other). Derrida will consistently interrogate Heidegger's characterization of animals as "poor in world," and he will spend a lot of page space and seminar time deconstructing the "as-structure" that Heidegger wants to reserve solely for human worlds (recall that humans are the only ones, Heidegger insists, who relate to the world *as* world, to a field of possibilities as such; while animals in Heidegger have some limited access to world, and to agency, they are not world-building or forming—they merely behave within their environment, they can't transform it, primarily because they have no access to the logos).[7] Countering this sense, Derrida will insist throughout his reading of Heidegger that being poor in world is still having some access to world and, thereby, to the privileges of the human realm.

It's crucial to recall, however, that Derrida's project in deconstructing the human/animal binary will *not* find its warrant in "granting" human privilege to animals—as if that were even possible—but will take the form of wondering whether humans are also *weltarm*, creatures poor in world who merely behave within an environment rather than exercising sovereign agency over it. In the end the haunting Derridean question will be whether both "animals" and "humans" share the fate constructed for animals in Western thinking: having no access to life or death "itself," we—like they—don't die but merely cease to exist. As Derrida writes about his "governing strategy" in discussing Heidegger and animality, "it would not simply consist in unfolding, multiplying, leafing through the structure of the 'as such,' or the opposition between 'as such' and 'not as such,' no more than it would consist in giving back to the animal what Heidegger says it is deprived of; it would obey the necessity of asking whether man, the human itself, has the 'as such'?" (*ATT* 159–60).

But even within this thoroughgoing critique of Heidegger, Derrida continues to accept (and in fact builds on) Heidegger's comparative method, defining *life* in terms of "world." To put it another way: the most obvious point of agreement between Heidegger and Derrida is their shared rejection of a biological or metaphysical thematization of life. Remember Heidegger from *World, Finitude, Solitude*: "the main thrust of our considerations does not rest upon a thematic metaphysics of life (of plants and animals)" (*WFS* 193); that "main thrust" will instead fall on the ontological question of various beings' actions within a world.

Derrida wants first and foremost to question Heidegger's confidence concerning humans' privileged and exclusive access to this thing called world (and to wonder whether there's ever any access to anything "as such"). But throughout Derrida's late work he continues to follow Heidegger in defining life (and death) through this overarching thematic of world. For example, when it comes time to publish the French edition of his many memorial-funeral pieces (originally collected in English in 2001 as *The Work of Mourning*), Derrida renames the collection *Each Time Unique, the End of the World* (*Chaque fois unique, la fin du monde* [2003]). In the "avant-propos" of *Chaque fois unique* Derrida lays out what he calls the "thesis" of the collection:

> The death of the other, not only but especially if one loves that other, does not announce an absence, a disappearance, the end of *this or that* life, that is to say, of the possibility of a world (always unique) to appear to a *given* living being. Death declares each time the end of the world in totality, the end of every possible world, and *each time the end of the world as unique totality, therefore irreplaceable and therefore infinite*. As if the *repetition* of the end of an infinite whole were once more possible: the end of the world *itself*, of the only world that exists, each time. Singularly. Irreversibly. For the other and in a strange way for the provisional survivor who endures this impossible experience. It is this that I would like to call "the world."[8]

In a characteristically Levinasian challenge to Heidegger's sense of world as Dasein's relation to the abstract possibilities of Being, the Derridean world is by contrast opened by the radical singularity of the other(s), a finite world of future possibility that is each time unique. Understood as opening to the necessary finitude or alterity that marks "our" lives, the world is haunted always by the end, insofar as the finite temporality of mortality is what makes "life" possible in the first place.

As Martin Hägglund explains in *Radical Atheism: Derrida and the Time of Life*, it is the Derridean "trace structure of time that is the condition for life in general. Whatever we do, we have always already said yes to the coming of the future, since without it nothing could happen."[9] If life is defined by a singular being's relations to future possibilities (a being that is alive is inexorably open to the future and, likewise, is necessarily open to mortality), then death constitutes the radical end of the world. At the end of a living thing's ability to endure through time, when it is no longer alive, that being does not escape time somehow but is left by death to live on (or not) through that very same trace structure (through, among other traces, the memories of the "provisional survivor"). Hägglund glosses this Derridean sense of "world": "The other is infinitely other—its alterity cannot be overcome or recuperated by anyone else— because the other is finite. . . . When someone dies it is not simply the end of someone who lives in the world; it is rather *the end of the world as such*, since each one is a singular and irretrievable origin of the world" (111). Hägglund argues at great and convincing length that the Derridean world is a site where the only definition of life is the life of finite beings (there is no God—no principle of untouched purity or everlastingness—or even the desirability thereof, hence the "radical atheism" of Hägglund's title). And when one of those singular finite beings is no more, the unique "world" that person had opened and inhabited ends as well.

Quite simply, if there is no other world, then death is each time the end of the whole world. Derrida eloquently sums this up in his eulogy for Louis Althusser: "What is coming to an end, what Louis is taking away with him, is not only something or other that we would have shared at some point or another, in one place or another, but the world itself, a certain origin of the world—his origin, no doubt, but also that of the world in which I lived, in which we lived a unique story. It is a world that is for us the whole world, the only world, and it sinks into an abyss from which no memory—even if we keep the memory, and we will keep it—can save it."[10] This likewise explains Derrida's obsessive interest in the line from Paul Celan's "Vast, Glowing Vault": "The world is gone, I must carry you."[11] After death we "live on" (or not) only as a series of traces in a series of other archives, other memories, other acts, other databases, other citations.

Importantly, Derrida's crucial work on animality extends these concepts of life, death, and world beyond the exclusively human realm (thereby posing his most serious question to Heidegger's analysis of death and world as exclusively the provenance of Dasein). For example, in the second year of the death penalty seminar (the session of 10 January 2001), Derrida insists that

> the death one makes or lets come . . . is not the end of this or that, this or that individual, the end of a who or a what *in the world*. Each time something dies [*ça*

meurt], it's the end of the world. Not the end of a world, but of the world, of the whole of the world, of the infinite opening of the world. And this is the case for no matter what living being, from the tree to the protozoa, from the mosquito to the human, death is infinite, it is the end of the infinite. The finitude of the infinite [*le fini de l'infini*].[12]

There is of course a lot that one could say about this astonishing Derridean opening of "world" and finitude beyond the human, and we will return to the question of Heidegger and the end of the world shortly.

But here I'd like to zero in on one very specific point: Derrida thematizes world as a question of life and death, a unique relation wherein a "living being" or a singular entity relates somehow to an "infinite opening" (world as a drama wherein "no matter what living being," human or otherwise, is dealing in some way with the "finitude of the infinite"). But given the definition of life that also gets configured here (as a singular entity's relation to the finitude of the world), it's not at all clear that most plants are technically among "the living" in Derrida's scheme of things. (This despite Derrida's inclusion here of the tree—as Deleuze and Guattari insist, that most arborescent and individuated plant form.) As Rodolphe Gasché has written about Hägglund's work, it's becoming clearer that "Derrida is essentially a philosopher of life, but of the only life there is—the life of finite beings."[13] Here, I don't particularly want to put pressure on the Derridean sense that temporality is the decisive defining factor for "finite beings" (the sense that living is essentially surviving—"living on" in relation to an indefinite, but mortal, future). Rather I'd like to emphasize the Derridean requirement that life and world accrue only to singular, each time unique, "beings." In short, if a kind of openness to the finite future within an individuated organism (a relation to world) is the ante to get in, it's not clear that many plants—especially those that are rhizomatic collectivities—can be said to be playing the game of life. In any case it seems clear that the question of what's alive and what's not in Derrida remains tied to a question of world, to futurity and at least some inkling of an individual (though not necessarily human) being's relation to mortality and futurity, "as such."[14]

Hägglund is the recent commentator who has done the most extensive work on Derrida and this question of life, though I would note for the interested reader that Richard Doyle's *On Beyond Living: Rhetorical Transformations of the Life Sciences* inaugurated this debate about life within Derrida scholarship twenty years ago. In a postpublication exchange, Hägglund was asked about the ambiguity surrounding "the question of the scope of 'life'" in *Radical*

Atheism—about the distinction between *différance* as a condition of (im)pos-
sibility for anything at all, and as the condition narrowly of living things and
their logic of survival. Hägglund responds:

> My answer is that everything in time is surviving, but not everything is alive. . . .
> The isotope that has a rate of radioactive decay across several billion years is in
> fact surviving, since it remains and disintegrates over time, but it is indifferent
> to its survival, since it is not alive. The two midges [that Derrida recalls from a
> fossil dated fifty million years ago], on the other hand, have a project, need, and
> desire. Like any other living being, they cannot be indifferent to their own sur-
> vival. This distinction is decisive for the definition of life in *Radical Atheism*. The
> reason I focus on life is because only with the advent of life is there desire in the
> universe. Survival is an unconditional condition for everything that is temporal,
> but only for a living being is the affirmation of survival unconditional, since
> only a living being cares about maintaining itself across an interval of time.[15]

In general this seems quite persuasive to me as a reading of Derrida: "survival [living on] is the unconditional condition for everything" in Derrida's
work, but not everything is "a living being": isotopes or volcanoes exist in and
change over time (they survive as entities), but they are not "alive" insofar as
they have nothing analogous to interests or desires (they have no phantasmatic world)—"since only a living being cares about maintaining itself across
an interval of time."[16]

So the pressure I'm interested in applying to the question of plant life in
Derrida does not primarily concern plants' existence in time, or the struggle
for continued survival as a condition for a definition of life.[17] Surely this drive
to live on is evidenced in plants, from the forest fight for sunlight to the roots'
competition for water or soil nutrients, or plants' agency in repelling predators
or attracting pollinators. But I'm wondering whether many species of plants (a
wall of ivy, a meadow of switchgrass, a pondful of lily pads—not to mention
plants' ancestors, the algae) can be thematized as a unique singular "being" or
an "itself"? A field of Kentucky bluegrass is more a collectivity (a smear of life
across series of emergent sites) than it is a singular, bounded living "being."
To put the question quite pointedly: on the grasslands that still cover more
than 30 percent of the earth's surface, are the individual blades of grass or daisy
stems "alive" in Derrida's sense? Do all plants have access to a phantasm of
futurity (or the lack thereof)? Does every single blade of grass have a "world,"
each time unique? When Sean Gaston writes, in *The Concept of World from*

Kant to Derrida, "this fiction of the world is an inescapable part of life and death, of animals and humans living and dying together. There is always the possibility that there is no world, but the world remains a necessary fiction,"[18] perhaps he puts the question most succinctly: does this "fiction of the world" shared by "animals and humans" extend to vegetable life?

My sense is that, in terms of this Derridean discussion of life and world, a wide swath of vegetable life remains *weltlos*—and thereby not technically "alive"—insofar as any given plant is not necessarily a singular entity and thereby cannot have a unique relation to futurity, or to the "each time unique" phantasm about (in)finitude that is the "world." If, as Timothy Clark argues in "What on World Is the Earth," the Derridean world is "something delusory in which one cannot not believe, simply by being alive,"[19] then it would seem that plants aren't alive, or at least they aren't alive in any similar way to humans or animals, because plants (as far as we know) do not or cannot project the fictional phantasm of a world. To put the question a different way: when Hägglund argues that living beings in Derrida "cannot be indifferent to their own survival" (to their own future), that's largely because they are *beings*, individual organisms whose survival or future can be thematized (however naively or phantasmatically) as "their own." Similarly, when Clark argues that world is something that "one cannot not believe," I might stress less the question of "believing" (or not) in the world and put more pressure on the question of the "one" who is doing the believing (or not). Surely, the desiring "one" remains as thoroughly phantasmatic as the possible "world" for Derrida, but the necessity for something like consciousness-projection or desire remains on both ends of this definition of "life." In short, if something is not such an individuated being (if it is not a phantasmatic site for living out a singular world of its ownmost desiring possibilities), then that thing is not, I think, technically "alive" on Derrida's terms.

Of course there are several caveats that immediately attach themselves to this seemingly perverse claim that plants are not alive for Derrida: first, plants are obviously "alive" in the everyday biological sense, for Derrida and for everyone else who treats these difficult questions, and I'd also have to admit that things are complicated considerably by Derrida's specific inclusion of the tree in his list of singular living beings with access to world. Things here get very murky indeed. On the one hand, a series of large oak trees, each one rising from an individual acorn, can I think easily be considered unique living beings, with a singular world of phantasmatic openness to other living things (this bird, squirrel, or set of humans), and be marked from the beginning by the eventual finitude of those possibilities; a stand of aspens, on the other hand, constitutes a rhizomatic collectivity (that's unique no doubt, but each individual tree is not really an "each one"). Maybe plants that reproduce sexually (through pollination and seed distribution) have "worlds" insofar as their offspring are "each time unique" beings (akin to a human or animal newborn), while plants that reproduce rhizomatically (underground through extending stolons) are *weltlos*. Here we might note that Derrida does at points offer childbirth as a privileged example of the event itself (absolute arrival that can't finally be calculated in a solely rational way), though his comments on cloning and "originary technicity" (the sense that we are always already technological copies rather than organic originals)[20] would, I think, complicate any wholesale privileging of sexual over asexual reproduction right out of the gate—not to mention the suspicious anthropomorphism inherent in granting a world only to things that reproduce in the manner that humans do.

We would also have to wonder at this point about Derrida's mention of the "world" of the protozoa, an example that I presume he chooses as much for its Greek etymology (*protos* + *zoon*, first life or first animal) as anything else. Of course, consideration of the microbial kingdoms opens up an even larger and more difficult series of questions for this discourse on life and world, and even more so for any discourse on the privileges of the unique human as the benchmark for life itself, insofar as there are ten "other" cells (literally trillions of microorganisms) for every "human" cell in each of our individual bodies.[21] In short, at the microlevel the "each time unique" human or animal entity begins to look like it's made up of a smear of other forms of life, most of which reproduce asexually. Looked at from the world of the protozoa, living bodies (such as human bodies) begin to seem less like individuated beings and more like a kind of prairie: a swarm of tangled and connected life-forms, with much of the life-activity taking place at a level unseen by the naked eye. In any case there are a lot of questions here, and I'll admit right now that I'm not sure I have any definitive answers concerning the liminal status of vegetable life in Derrida.

However, as I'll try to suggest in the rest of this chapter, this Derridean swerve around plants in relation to life is not merely evidence that Derrida callously abjects or ignores the vegetable kingdom. Rather, the excluded question of "plants, flowers, and vegetables" in Derrida becomes transversally attached less to the discourse of animal life than to the thematics of *physis* itself (often translated as "nature" but tending to have the sense in Greek of that "power of growth" linked to plant life, the condition of possibility for all life). Like *différance, physis* consists of a power of emergence that remains indifferent to this or that form of "life," to this or that individual being or its world: a necessary condition of (im)possibility that (in the end, or from the beginning) is wholly indifferent to any individual stone, animal, plant, or human.

To be clear, then: I'm not interested here in accusing Derrida of zoocentrism in his work on animal life. At the same time, though, one would be remiss in not wondering about the uncomfortable silence on the question of vegetative life both within Derrida's recounting of philosophy's shabby treatment of animals and, even more so, in the wider flood of recent work on biopower and animality—where Derrida's work has found a new and amenable home. (As I noted in my Preface and will discuss at more length in the Coda, Cary Wolfe—without a doubt the most prominent animal studies theorist—roundly rejects plants' claims to the life-conversation at the end of his *Before the Law*.) And you can't lay this exclusion of plants at philosophy's door, either—insofar as Aristotle's *De anima*, the founding Western text on "life," contains a plethora of thinking about plant life and its relations to human and animal life. Indeed, everyone in animal studies likes to recall Aristotle's sense of humans as "rational animals," but few recall that for Aristotle we are also "walking plants," as we saw in Chapter 2. In any case I'm *following* Derrida rather than *critiquing* him here—tracking him like the animal that therefore he is. But in doing so, I'm also of course following him like a plant—tracing the innumerable seeds he's disseminated from the flower of deconstruction.

To lay my cards fully on the table, I want to suggest that abundant resources for thinking about plant life in Derrida's work are to be found in his (now, it seems, all but forgotten) 1974 masterwork *Glas*. Among the many, many other debates that *Glas* engages (Hegel versus Genet, the law versus the outlaw, philosophy versus literature, normativity versus queerness, proper versus improper), I want to linger here on *Glas*'s staging of the animal versus the plant—life understood as animality versus life understood as vegetality; animal reproduction as appropriation versus the flower's expropriative deluge of pollen or nectar; the animal religions of ennobling sacrifice (from Judaism's myriad animal sacrifices all the way to Christianity's Lamb of God) squared off against the flower religions of excessive expenditure (from the murderous Cain, whose fruits of the soil displeased God, to the ancient flower religions of India and the cults of the Bacchian grape). I'll even go so far as suggesting that Derrida's final words in his final seminar (*The Beast and the Sovereign*, vol. 2) turn not on the animal and the question of "higher" life-forms but turn instead back toward the vegetable "world" of uncontrolled, cancerous growth and indifference. In the end Derrida turns not to the individuated animality of bios or zoe but to vegetable physis as the inhuman power of emergence or event. Perhaps, in the end, the plant (as the privileged figure for growth or emergence itself) doesn't have access to world because it embodies and deploys something prior to (indeed, superior to) the problem of world: the Plant and (or maybe even as) the Sovereign?

If nothing else, I want to demonstrate that the notable absence of deconstructive thinking about vegetal life in Derrida's late work is more than made up for in *Glas*, which performs what one might call a classic deconstruction of the animal/plant opposition. *Glas*'s thoroughgoing investigation of what Derrida names "the question of the plant, of *phuein*, of nature" will lead him to posit straightforwardly that "practical deconstruction of the transcendental effect is at work in the structure of the flower" (*G* 15b). And here I'm likewise following what Michael Naas in *Derrida from Now On* has wryly called "an almost failsafe hermeneutic principle when trying to check any hypothesis or thesis regarding the work of Jacques Derrida. The principle runs something like this: make your case by ranging widely through Derrida's corpus . . . but *then* turn at the end of the day to *Glas* to see whether the whole thing was not already laid out for you, from start to finish, in 1974."[22]

Glas

As John Leavey, one of the English translators of *Glas*, ironically noted in his 2012 Modern Language Association conference presentation, Derrida's experimental 1974 text is a consistent winner in the academic "best book you've never read" game that sometimes breaks out after several glasses of wine at college-town dinner parties. Thereby if nothing else, the obscurity of *Glas* constitutes an ironic bookend to Derrida's immensely popular work on animality. And one should be quick to point out that *Glas* is a justifiably infamous monster—consisting of two columns on each (extrawide) page of the text, one concerning Hegel, the other Jean Genet. And of course to say "it's written in two columns" seriously underestimates the difficulties that *Glas* presents, as Derrida complicates his own binary schema by consistently inserting other texts within the columns ("tattoos" of different-fonted text in boxes, along with citations, dictionary definitions, and other breaks where myriad voices or texts intervene).

Whatever else it is, *Glas* is a tour de force comedy (and the genre is literally comic, a circle—the beginning is the end, *Finnegans Wake* style). And the joke is aimed squarely at that most serious and tragic of philosophical projects, Hegel's. The last word of the Hegel column (such as it has a "last word") is literally "comedy."[23] In a world that has posthumously decided Derrida is either a thinker of animals or a mournful thinker of "prayers and tears," *Glas* intensely recalls for us another Derrida, the philosopher of malicious performative joy and relentlessly avant-garde artistic production. And the Derrida who's a thinker of vegetable life. As Claudette Sartiliot notes in *Herbarium/Verbarium: The Discourse of Flowers*, "when Derrida produces in *Glas* a cross-fertilization of philosophy and literature, Hegel and Genet, we should recognize his work not only as critical and aesthetic, but as biological."[24]

Readers well-versed in *Glas* will forgive me if I lay out a brief introductory roadmap for those unfamiliar with the text. First, and most obvious, the text's two columns are set up to test the Hegelian laws of dialectical sublation— Hegel's famous *Aufhebung*, the engine of dialectical progress: affirmation, negation, synthesis. The most obvious question posed by the text (as well as performed in any attempt to "read" it) concerns the first column, which takes up the project of dialectical progress posed by Hegel—"a higher calculus without remains, what consciousness wants to be" (*G* 60a). The question that inaugurates the book, and "remains" throughout it, is whether this consciousness can, as Hegel would lead us to believe, "sublate" column b (which in this case happens to concern the proper name for impropriety itself, the literary project of transgression and masochistic queer desire performed in Jean Genet's literary writings). As a voice (maybe Derrida's?) puts it some sixty-five pages into the text, "Why make a knife pass between two texts? Why, at least, write two texts at once? What scene is being played? What is desired? In other words, what is there to be afraid of? Who is afraid? Of whom? There is a wish to make writing ungraspable, of course. When your head is full of the matters here, you are reminded that the law of the text is in the other, and so on endlessly. . . . You are no longer let know where the head of the discourse is, or the body" (*G* 64–65b). While reading *Glas* seems far from the ordinary experience of responding to texts, Derrida's gambit will be to show that such multiplying intertextuality is the most basic practice of all reading and writing: Even "if I line myself up and believe—silliness—that I write only one text at a time, that comes back to the same thing, and the cost of the margin must still be reckoned with" (*G* 66b); and the text demonstrates this productive undecidability in a wide variety of discursive regimes. Indeed, *Glas* moves like a plant or a vine, by "agglutinating rather than demonstrating" (*G* 75b), by cross-pollinating, becoming invasive, coiling (*G* 104–5b), and, above all, by what Derrida calls (vegetable) grafting or cutting of one text and hybridizing it with another (*G* 108a&b). Perhaps the only thing *Glas* doesn't allow is dialectical sublation—and Derrida will go on to suggest that this is in fact the one and only law of the dialectic: its condition of possibility is simultaneously its condition of impossibility.

Of course, for Hegel a certain kind of undecidability ("tarrying with the negative") is absolutely necessary for the progress of spirit—which must constantly risk knowledge and life by confronting (and overcoming) ignorance and death. This is the dialectic in a nutshell, negating difference and raising it (however momentarily) to a higher level of synthesis. The dialectic teaches, if nothing else, how you make the negative productive—with death, for example: you need to take death within the itinerary of your life, tarry with it and raise it up, make it into an engine for progress and discovery rather than treating death's negativity as a sinkhole of despair. Otherwise, death simply defeats you. This is why, in his reading of Hegel's master/slave dialectic, Jacques Lacan suggests that death is "the absolute master": the slave, because he fears and knows death, can in fact work, transform his circumstances, can use the absolute otherness of nonexistence to turn a temporary profit through work. The otherness confronted by the Hegelian negative is then not paralyzing but is what consciousness requires to become stronger, more confident, to progress: consciousness must feed on the remains each time an opposition is adjudicated, reappropriating these remains, raising them up. If Hegel is often smeared as a totalizing thinker, it's important to note that his vision of the absolute is not a static state but endless movement: absolute spirit is perhaps nothing other than the command that you have constantly to lose yourself in order to find yourself—endlessly risk death in order to overcome it.

Derrida demonstrates in *Glas* that Hegel is a thinker of animal desire as the engine of knowledge and progress, but this animal desire must be negated and raised—made culturally respectable by spirit. As Derrida writes, in Hegel "human feeling is still animal. The animal limitation, I feel it as spirit, like a negative constraint from which I try to free myself, a lack I try to fill up. . . . Man passes from feeling to conceiving only by suppressing the pressure, what the animal, according to Hegel, could not do" (*G* 25a). And of course this Hegelian tension between "the animal moment and the spiritual moment of life" (*G* 25a) begins a biopolitical project that becomes very familiar to us throughout the

nineteenth century to our own day: the project of sorting out man's animal desire and refining it (or not) through education, repression, or normalization
by culture. As we saw in Chapter 1, the theme of man's subtending animality
runs straight through all the thinkers who usher in what we like to think of as
"our" biopolitical era. In short, Derrida's reading of Hegel comes to a conclusion similar to Foucault's in *The Order of Things*: modernity is not, as some
animal studies thinkers would have us believe, born by jettisoning or abjecting
animality but rather by fully incorporating animal desire into our definition of
the human.

And for his part Derrida doesn't jettison this "animal" Hegelian project at
all but in fact intensifies it, puts it to its own test: does or will this animal appetite for appropriating otherness succeed in the end? Can the negative be so
easily economized, made to pay off as knowledge at a higher level? As Derrida
sums up the obsessions of *Glas*, "What remains irresoluble, impracticable, non-
normal, or nonnormalizable is what interests and constrains us here" (*G* 5a):
the text lingers over all those remains that refuse to be lifted up, normalized,
incorporated, known.

But *Glas* does not merely stage for us a Manichean struggle of norms versus excess, with some flavor of excess as the inevitable winner. This is a popular rendering of the project of deconstruction, but nowhere is it clearer than
in *Glas* that Derrida does not simply celebrate a negative undecidability—the
flower, literature, queer desire, the mother, the sister, etc. It's not that these "un-
decidables" are simply the other face of Hegel's project of accounting for everything, but they constitute their own (for Derrida, oddly "superior") form of
totalization, a form that can in fact account for its own necessary failure, for
its own remains, in a positive rather than negative way. In short, the drama
that *Glas* stages for us is not merely Hegel versus Genet (totalization versus
undecidability, philosophy versus literature, animal versus plant), but the book
performs two differing modes of totalization, what Derrida calls two different
ways of "saying everything."[25]

Genet's commitment to the "flowery" language of literature, in other words,
is not deployed merely to subvert or impede the totalizing claims of Hegel's
philosophical language (it's not just an inert cinderblock thrown onto the high-
way of spirit), but the Genet-literature-queer-flower series forms a robust alternative mode of thinking about everything. As Derrida writes about the relations
between literature and philosophy throughout his career, "My 'first' inclination
wasn't really toward philosophy, but rather towards literature—no, towards

something that literature accommodates more easily than philosophy."²⁶ In other words it's not so much that literature rests in a simple binary opposition to philosophy in Derrida's thought, but philosophy and literature are two projects or performative idioms that enact two distinct modes of totalization. It's not that literature (and, with it, the entire Genet column of *Glas*) merely wallows in being open-ended and indeterminate, while philosophy thirsts after rationalist totalization; rather, both literature and philosophy embrace their own modes of totalization, differing attempts to say everything. What separates them is not their nature, or even their project, but the performative and idiomatic ways in which literature "accommodates more easily than philosophy" the necessary failure of the project of saying everything—its inevitable limits, as well as its productivities.

There is a sense in which Derrida wants to treat philosophical works in a "literary" manner, but this is only to say that Derrida is drawn in the text of Western philosophy to those moments of hesitation, aporia, uncertainty: where the texts of the otherwise confident philosophical tradition fold back on themselves, those moments when they begin to accommodate themselves or bargain with a project of accounting for everything that shows itself simultaneously as an impossibility. Literature, if there is such a thing, is the proper name for that (im)possibility of totalization enacted as a mode of writing: literature constitutes a singular idiom or form of engagement (different each time) that is rare in its ability to move forward in the project of saying everything while simultaneously acknowledging its own limitations. At the end of this chapter we'll return, along with Derrida, to this question of totalization, of everything.

In any case Derrida incessantly tests Hegel's logic of raising the remains by contrasting him with Genet on a number of topics—from death and burial (literally, normativizing dangerous remains—making mourning out of melancholia) through the question of the family (which is at base the normalization of an otherwise dangerous sexual desire, raising it to the level of propriety—though there is much discussion of Hegel's bastard son and his odd attraction to his crazy sister). But I'm going to argue that the privileged thread through Derrida's interrogation is the question of *animal* desire in Hegel (whose dialectic is nothing if not the animal movement of appropriative desire) versus the masochistic, un(re)productive queer desire of Genet (whose very name is a *plant*, not to mention his most famous book, *Our Lady of the Flowers*). Perhaps the privileged binary clash within *Glas*, in other words, is less philosophy versus literature than it is animal versus plant.

Plants and Animals in *Glas*

At the beginning of the Hegel column Derrida very economically focuses the Hegelian corpus on two moments that he wants to interrogate: "two very determined, partial, and particular passages, two examples. But perhaps the example trifles with the essence": "First passage: the religion of flowers. . . . Second passage: the phallic column of India" (*G* 2a). Why, one wonders, these two strange moments in Hegel's texts, the first from the *Phenomenology* and the second from the *Aesthetics*? How can the religion of flowers and the Dionysiac celebration of fertility cults (practices recalled from ancient India and Africa) be privileged moments to begin thinking about modern Europe's most ambitious Enlightenment thinker? Initially, one might begin to answer the question "why would Derrida begin here?" by venturing that these are both originary, or maybe even preoriginary, moments in Hegel. Everything in Hegel moves in threes, of course (affirmation, negation, synthesis), and his sweeping eagle-eye views of history and religion are no exceptions: like the opening of the *Phenomenology* itself, which moves from sense-certainty, through unhappy consciousness, to the first inklings of something like knowledge, history for Hegel moves from the ancient world (and its naive "sense-certainty" religions of flower innocence and fertile desire) through the Judeo-Early Christian world of guilt and culpability, toward their dialectical sublation in the Christian Enlightenment under way in Hegel's own day.

So the religion of flowers or the fertility cults of the ancient world are less originary moments in Hegel than they are preoriginary moments, naive non-starters. Like sense-certainty (the necessary but by no means sufficient condition for perception), the religion of flowers constitutes a historical phase that exists in Hegel's system only to be overcome. As Derrida writes, "Flower religion [in Hegel] is not even a moment or station. It all but exhausts itself in passage, a disappearing movement, the effluvium floating above a procession, the march from innocence to guilt. Flower religion would be innocent, animal religion culpable. Flower religion . . . no longer, or hardly, remains; it proceeds to its own placement in culpability, its very own animalization, to innocence becoming culpable and thus serious" (*G* 2a). Derrida then quotes Hegel from the *Phenomenology*: "The innocence of the flower-religion, which is merely self-less representation of the self, passes into the seriousness of warring life, into the guilt of animal religions; the quiet and impotence of contemplative individuality pass into destructive being-for-self" (*G* 2a), and thereby into the dramas of progress, history, law, and politics. Like the places from whence they came (Africa and India), flower religions are literally prehistorical for Hegel; they are merely a "quiet and impotent" state of nature that has yet to be negated, yet to pass over into the dialectical realm that is culture. As Derrida quotes Hegel later on: "the purpose of nature is to kill itself and break through its shell of the immediate, of the sensible, to consume itself like a Phoenix, in order to upsurge, rejuvenated, from this exteriority, as spirit" (G 117a). Nature must be raised and transformed; nature must become culture.

And let's recall that a certain resurgence of innocent flower religion (going by the modern name of "romanticism") is one of Hegel's primary targets in the *Phenomenology*; most specifically, he takes aim at Goethe's and Schelling's nature-obsessed romanticism, "a view which is in our time as prevalent as it is pretentious,"[27] Hegel snickers. Functioning as a naive, "bad infinity"—nature as preoriginary indifference, unable to account for or direct progress—this new-fangled flower religion is famously mocked by Hegel as "the night in which all cows are black."[28]

And yet, Derrida asks, and yet. . . . Just as sense certainty will be difficult to leave behind on consciousness's ascent toward knowledge, one might summarize the whole of *Glas* as an extended question posed to Hegel's discourse: if Hegel's thinking of culture and law is a graduated process of extracting cultural sacrifice, progress, and knowledge from an indifferent, asignifying nature—animalizing the plant, so to speak, and then humanizing the animal—will this project really ever be able to gain and maintain escape velocity from its primary other and condition of possibility, a kind of preoriginary *physis*? The question remains: can the dialectic "escape" or sublate the otherness on which it depends for its fuel: nature, *physis*, the plant, growth, and emergence—*this preoriginary indifference that (secretly) calls for and drives the law*? Can you "train" the event of emergence with desire-as-lack, shame, or propriety? Or is the event of emergence indifferent to culture, just not the kind of thing that will tamely undergo such higher-purpose "progress"? And doesn't such a system of cultural differentiation require (endless?) production of ever-more indifferent remains for the whole system to be able to move to a higher level? More remains, more desire, higher plateaus for spirit: "By removing itself from nature, by denying nature within itself, by relieving [sublating], sublimating, idealizing itself, desire becomes more and more desiring. Thus human desire is more desiring than animal desire; masculine desire is more desiring than feminine desire, which remains closer to nature. More desiring, it is then more unsatisfied and more insatiable" (*G* 169a). As nature must be cultured, so the plant must be animal-ized: "The more one is raised in(to) the differentiating hierarchy of animalness, . . . the more the organism is capable of assimilating foreign bodies or differen-tiated organic totalities" (*G* 114a).

To fast-forward through all the avatars of plant religion in *Glas*—literature, the queer, the convict, the erection, the crazy sister, the illegitimate child, all sorts of nonsignifying desire—the question for all of them will be similar: can these "preoriginary," nonnormative states finally be normativized, animalized, economized, made to pay, made to progress toward higher knowledge? Can the inert or indifferent condition of possibility for dialectical sublation merely be taken up into the movement of knowledge, negated and lifted to the absolute by the engine of spirit?

At the macro level *Glas* stages for us a face-off between the animality of appropriating, civilizing movement of spirit and the "law" of plant-nature—which Hegel will grant is growth yes, emergence yes, but a kind of undifferenti-ated and uncontrolled growth that is finally anathema to the progress of spirit or law—and not because nature or the plant (or literature, for that matter) is merely inert but because they're dangerous. The "bad infinite" plant (which, Aristotle reminds us, has no *psukhe* or power of its own, other than growth without telos) must be schooled by the animal in the "good infinite" of cul-pability, desire for objects that can be worked on by spirit. Animal desire is "superior" for Hegel because that desire can be repressed, sublated, trained. Plant desire merely grows without telos, without a proper end (or a proper death)—and the logic here is classic: in Aristotle, as we've seen, the vegetable *psukhe* or soul deploys a single power, the ability to grow. But that power is without entelechy—plants have no higher, ideal form toward which they de-velop. They just grow uncontrollably until they grow no more.

So in this vein, for example, Derrida will recall for us Hegel's account of the feminine, and its philosophical prejudices concerning the woman and the plant: "This discourse on sexual difference belongs to the philosophy of nature. It concerns the natural life of differentiated animals. Silent about the lower animals and about the limit that determines them, this discourse also excludes plants. There would be no sexual difference in plants. . . . In this sense, the human female, who has not yet developed the difference or opposi-tion, holds herself nearer the plant" (*G* 114a). Or as Hegel himself puts it, "The difference between man and woman is that of animal and plant" (quoted in *G* 191a). So for Hegel the plant must be "animalized" in the same way that the woman must be manned, the family must be nationed, and mere matter must be lifted up by spirit: without that sublating moment there's nothing but cancerous "natural" growth, without regard to betterment or higher ends: nutritive life without a "world."

Again it is the plant, which Derrida consistently links to the question of *physis*, that remains the linchpin site of inquiry in *Glas*. Not the animal, which Derrida demonstrates time and again is already incorporated into the movement of spirit by Hegel (animality may in fact be said to be the ground of that movement of desire). Hegel's animal desire is the engine of the dialectical discourse, whereas the "plant desire" endlessly on display in the Genet column doesn't "lead" anywhere, or anywhere proper. Masturbation, queer desire, prison, the thief's underworld, cathection onto the mother: can any of these be raised? Like a flower, and connected centrally to the erections that populate Genet's text like flowers, this "other" desire emerges intensely and lingers for a moment before it releases its seed to the wind, and it's gone. It falls, to the remains, as remains, downward as it were, back to the bad-infinite ground—as opposed to those remains constituting an engine for higher knowledge. The Genet column of *Glas* resolutely resists that raising, remaining with non-reproductive desire, with sense-certainty, with the plant: *physis*. In a crowning irony, on Hegel's account, Genet's queer desire is shown finally to be "natural desire"—the plant, the flower, the queer outlaw, all remaining indifferent to culture, to "raising." And Derrida clearly demonstrates this in the Genet column through the figure of the erection, that privileged sign for male sovereignty and reproductive power, that is of course "raised" by means that are largely out of any subject's rational control. In the Genet column erections grow like flowers and spill their seed all over the place, but their activity doesn't lead to anything that Hegel would recognize as a "higher" state.

And of course this nonnormative activity is all the more necessary because cultural progress depends on a remainder that it finally cannot control: any progressive dialectic must consistently produce ever more new, indifferent, asignifying remainders—emergent undocumented material that needs to be rendered meaningful, negated, and raised. Derrida summarizes *Glas*'s reading of the Hegelian dialectic in concise fashion (in a passage that Derrida pointed out in the *Glas* seminar that I took from him):

> There is no choosing here: each time a discourse contra the transcendental
> is held, a matrix—the (con)striction itself—constrains the discourse to place
> nontranscendental, the outside of the transcendental field, the excluded, in the

structuring position. The matrix in question constitutes the excluded as transcendental of the transcendental, as imitation transcendental, transcendental contraband. The contraband is not-yet dialectical contradiction. To be sure, the contraband necessarily becomes that, but its not-yet is not-yet the teleological anticipation, which results in its never becoming dialectical contradiction. The contraband remains something other than what, necessarily, it is to become. Such would be the (nondialectical) law of the (dialectical) stricture. (G 244a)

Within this general law of the dialectic the emergence of difference can only look like indifference, a remainder that is not-yet subject to a law, made whole. But of course the emergence of this not-yet, this indifferent difference, is at the end of the day the structural principle of the entire discourse. The dialectic is stale as last night's beer without the emergence of the event, the remainder, the contraband. The "excluded" (the bad infinity of supposedly indifferent nature) finds itself in fact "in the structuring position," "the (nondialectical) law of the (dialectical) structure." The structure whereby this remainder is ineluctably produced by dialectical sublation is most concisely worked out in *Glas* according to the logic of the plant, though again it's worth remembering that the plant is linked by Derrida directly to the *phuein* of *physis*—the power of growth and emergence that is in turn linked with "nature" in most Western thinking: "the flower appears in its disappearance, vacillates like all the representative mediations, but also excludes itself from oppositional structure" (G 246a).

When summing up the stakes and movements of *Glas* late in the text, Derrida asks straightforwardly, "How far have we got?" In answering that question, he returns to his opening focus on plant and animal in Hegel:

> Why, plant and animal, plant then animal? . . . No opposition can form itself without beginning to interiorize itself. This organicity already binds itself again to itself in the plant, but life represents itself therein only by anticipation. The actual war, as opposition internal to the living, is not yet unchained. The plant, as such, lives in peace: substance, to be sure, and there was not yet any substantiality in the light, but peaceful substance, without this inner war that characterizes animality. Already life and self, but not yet the war of desire. Life without desire—the plant is a kind of sister. (G 245a)

In the end Derrida suggests that Hegel elides plant life because "the subjectivity of the plant is not yet for itself. The criterion for this is classic: the plant does not give itself its own place" (G 245a).

From *Glas* to the Final Word

To return from *Glas* to a logic of Derrida's later work on animality, it's hard not to conclude that Derrida's analysis of the plant in Hegel recoils immediately on Hegel's central theme, the animal privileges of human consciousness. Recall that Derrida's work on animality is not offered in the name of lifting up the animal (suggesting the animal is similar to the human in sensation, suffering, death, relation to world, etc.) but rather wonders whether all the things that Western thinking has said about the animal (it behaves rather than acts, it has no access to entities "as such," it has no knowledge of finitude) are finally all that can be said about humans as well.

In *Glas* this seems clearly the gambit and logic of Derrida's obsession with the plant as well: are humans ever lifted to the level of being "for themselves," giving themselves their own place? Or are humans somehow akin to plants—at the end of the day subject wholly to their environments, living, growing without entelechy, and dying indifferently: responding as they can to the events that happen around them, events or emergences within a physical world largely indifferent to any particular life-form "as such"? In short, *Glas* asks us: can what Hegel calls the "innocent indifference of plant life" finally be incorporated, mastered, and left behind by the differentiating work of spirit, "this determinateness and this negativity" (*G* 246a)? Does human-animal desire triumph over vegetable *physis* in the end? Or is it the other way around? Suggesting again that *physis* is less "innocent" than it is "sovereign."

In fact, one is left to wonder whether Derrida's work on plant life in *Glas* finally ups the ante around the question of life, even more so than the work on animality—insofar as *Glas* suggests that a kind of plant logic is "superior" to the animal logic of Hegel's humanism. (Hence the upped ante—Derrida's work on animals demonstrates that the logic of animality may in fact be the logic of humanism, but to my knowledge he never suggests that the logic of animality is somehow superior to the logic of humanism.)

I think the privileged place to go from here in exploring the upshot of this question is right to the very end of Derrida's itinerary, to his final lecture course before his untimely death from pancreatic cancer: the second series of *The Beast and the Sovereign* lectures, where he both reopens and is tragically forced to end his long engagement with Heidegger and the question of animality. This last lecture course is unique, as Michael Naas has definitively shown in *The End of the World and Other Teachable Moments*, insofar as it's marked by a discontinuous but obsessive investigation of a single word that emerges as key in Heidegger's texts of the 1930s, the German word *Walten* (as a kind of sovereign, originary violence). As Derrida puts it in the final lecture of what turned out to be his final seminar, "late in my life of reading Heidegger, I have just discovered a word that seems to oblige me to put everything in a new perspective" (*BS* 2:279). As he summarizes what he takes from Heidegger's work on animality in the second volume of *The Beast and the Sovereign*, "I hang onto this curious non-sequitur that consists in defining animality by life, life by the possibility of death, and yet, and yet, in denying dying properly speaking to the animal. . . . What is lacking is not supposedly access to the entity, but access to the entity *as such*, i.e. that slight difference between Being and beings that, as we shall see, springs from what can only be called a certain *Walten*" (*BS* 2:116).

In his last lecture course Derrida returns one final time to Heidegger's work on animality and world, recalling for us Heidegger's original gambit: to think life not as an essence, property, or thing but to think various forms of life always within a series of relations to their origins and possibilities, to *physis* and world. Following a fundamental ontology, Heidegger insists that we must return to *physis* in order to rethink what human life is: as Derrida reminds us, "Remember that for Heidegger—because we'll need to think seriously about this—*physis* is not yet objective nature but the whole of the originary world in its appearing and its originary growing. It is toward this originary 'world,' this *physis* older than the objective nature of the natural sciences, . . . that we must turn our thought in order to speak anew and differently about the being-in-the-world of man or of Dasein and animals, of their differential relation to this world that is supposed to be both common and not common to them" (*BS* 2:12–13).

Throughout this final lecture course Derrida tracks Heidegger's incessant use of the word *Walten* and its variants within the conversation about the questions of animality, world, and death, suggesting that for Heidegger there is a kind of originary granting even before the "world" of Dasein's possibility—or at least suggesting that what's gifted in Heidegger's notion of *physis* is tinged less with Hegel's preoriginary "innocence" than with a certain kind of originary violence: as Derrida writes, "*walten*: always the force of this same word that bespeaks a force, a power, a dominance, even a sovereignty unlike any other ['more sovereign than all sovereignty,' he says in a footnote]—whence the difficulty that we have in thinking it, determining it, and of course translating it" (*BS* 2:123).[29] Derrida finally ventures this translation of *Walten* in Heidegger: "a sovereignty of the last instance, . . . a superpower that decides everything in the first and last instance" (*BS* 2:278). And Derrida consistently notes the link-age of *Walten* to the question of *physis* in Heidegger: "*Physis* means this whole *Walten* that prevails through man himself and over which he has no power, of which he is not the master. . . . In other words, this all-powerful sovereignty of *Walten* is neither solely political nor solely theological. It therefore exceeds and precedes the theologico-political" (*BS* 2:41). Derrida goes on to defend "my insistence on *Walten* here, as a figure of almost absolute power, of sovereignty before even its political determination" through its originary links to *physis*: "Heidegger explains to us that if we translate more intelligently and clearly, if we (that is, he) translate *physis* not so much by growth (*Wachstum*) as by *Walten* ['the self-constituting, self-formed, sovereign predominance of beings in their totality'], if, then, we translate *physis* by *Walten* rather than *Wachstum* (as sovereign power rather than growth), that is, as Heidegger expressly says, because it is clearer and closer to the originary sense, the intentional sense, the meaning of the originary sense or the originary meaning of *physis*" (*BS* 2:40).

In passing, we might note that this emphasis on an originary *Walten*-as-*physis* helps explain the otherwise very odd interplay in *The Beast and the Sovereign* lectures between Derrida's readings of *Robinson Crusoe* and Heidegger's lectures in *World, Finitude, Solitude*—around the questions of access, animal-ity, solitude, and world, and their relations to an even more primordial *physis*, glossed by Derrida as "what increases or grows, growing, growth, the very thing that has grown in such a growth. . . . And there too, the pedagogy by which Heidegger illustrates what he means by 'growing,' by growing as nature, as realm or domination of *physis*, takes the form of a Robinsonian landscape: the plants, the animals, the seasons, the day and the night, the stars, the tem-pest and the storm, the raging elements" (*BS* 2:39). This is at least partially to say that here, in the very last lecture course, Derrida returns to the "*Glas*-struc-ture" of literature before philosophy and the linkage of the literary to the logic of the flower. Philosophy's insistence on "world" is pitted against literature's in-sistence that, as Derrida loosely translates Celan by channeling Crusoe, "there is no world, there are only islands" (*BS* 2:9): in the sense, perhaps, there is no such thing as "the novel," there are only novels; there is no "animal in general," only particular animals; there is no such thing as life, only living things. This likewise helps explain why Derrida characterizes death quite literally, as "each time unique, the end of the world."

In any case we follow Derrida in circling back to *Walten*—literally the last word of Derrida's final lecture course—which concerns the sovereign, originary reign of *physis*—not in terms of an animal or human access to world but in terms of a "vegetable" power of growth or emergence that's indifferent to man or animal and thereby not so much animating or reassuring but crushing in its indifference. At the very end of the day one might say that *physis*-as-*Walten* may in fact constitute the ironic realization of Hegel's all-encompassing dream for philosophy: an originary force that in fact constitutes and lords over everything, all the rest. But that force turns out to be *physis*, not *Bildung*; vegetable, not animal; nature, not culture—the condition of possibility for a life, but not alive.

In the very end there undoubtedly will be a kind of totalization waiting for the human—as there has been for the 99.9 percent of all living species that have already gone extinct over the history of this planet. But that sovereign totalization happens *to* us, not *for* us. And Derrida wraps up his analysis of Heidegger on precisely this point: "the as, the as-structure [of access to world] that distinguishes man from the animal is thus indeed what the violence of *Walten* makes possible" (*BS* 2:288). The human-animal "as-structure" of world, in other words, depends wholly on a prior field of indifferent forces, "a superpower that decides everything in the first and last instance." Just as humans violently displaced *physis* to make way for the Anthropocene, that very same *Walten*-as-*physis* will—in the "last instance"—violently displace and overtake each and every one of us.

So we come again to *Glas* and the serene indifference of the plant as the most intense figure for the power of *physis*—all the more a superior logic for its originary, sovereign force of emergence and growth, coupled with its utter indifference to this or that living thing. Like the deep time of *différance*, the Derridean plant doesn't *have access* to a past or a future world; but maybe that's only because the originary *Walten* of vegetal *physis* comprises both the absolute past and the posthuman future. An intense irruption of *physis*, the plant as such is not technically alive in Derrida (it has no phantasmatic world or singularity), but it is the condition of (im)possibility for "life." To circle back to *Glas*, it shows us in the end that the supposed privileges of human-animal desiring life arise by necessity out of a nonliving principle—*physis*, *différance*, vegetable life, call it what you will—that afflicts and marks the supposed living presence of *Geist* or world: emergence happens in time, which is not to be confused with the phantasms of human temporality. And whatever emerges—living or otherwise—is by necessity afflicted by this trace structure (infected by its conditions of emergence).

Correlationism and Other Phantasms

If nothing else, this kind of analysis helps us to understand a bit more clearly how "life" works in Derrida—what participates in the discourse of life (strictly speaking, singular "beings" with some relation to futurity) and what doesn't (presumably, everything else); or, to put it somewhat differently, this analysis confirms that Derrida is certainly a thinker of life, but he's not a vitalist: which is to say, everything is subject to the law of the trace (of time and finitude), but not everything is "alive" in Derrida's philosophy. And I think this question of "life" and its discontents remains an important one to explore because it answers the growing consensus that, while Derrida's work on animality shows that his thought is not, strictly speaking, anthropomorphic, Derrida nevertheless remains a secretly human-subject-centered, or "correlationist," thinker. The word *correlationism* was coined by Quentin Meillassoux in *After Finitude* as an attempt to get back behind the Kantian Copernican revolution of subjectivity, and Meillassoux defines the term quite simply: "Correlationism consists in disqualifying the claim that it is possible to consider the realms of subjectivity and objectivity independently of one another."[30] For Meillassoux, and a series of other "speculative realist" thinkers, it's the great "outside" of the real that was lost with Kant, then further mystified by Hegel, bracketed by Husserl, and ignored as "essentialist" or "metaphysical" by virtually all continental existentialism and phenomenology. These modern and postmodern discourses remain (on Meillassoux's account) hopelessly filtered through (and thereby openly centered on) human subjectivity, precisely insofar as these philosophies all follow Kant in resolutely refusing to say anything at all about the real, or at least not without saying something at the same time about the perceiving subject. And it is precisely this sense (accounting always for the each-time-unique human subject—for its language, idiom, finitude, and history) in which deconstruction has been painted as the correlationist par excellence. As Slavoj Žižek puts it, "If there is a philosopher who effectively seems to be caught in the circle of what Meillassoux calls 'failed correlationism,' it is Derrida."[31]

As a riposte to phenomenology and deconstruction, then, the dream of anticorrelationist discourse is not to continue endlessly rethinking the relations between humans and their world (the relation between the logos and the "as such" of Being) or even to extend that relational axiomatics to animals or other forms of life (which characterizes the vitalist project); rather, the project is to speculate again about the nature of reality itself, independent of humans' perception of it. The goal is to describe "the nature of being, rather than the human philosopher's approach to it" (31), as Ian Bogost puts it in *Alien Phenomenology*.[32] Or as Levi Bryant, Nick Srnicek, and Graham Harman argue in their introduction to *The Speculative Turn*, the project consists in getting back behind phenomenology, with its obsessive focus on human subjectivity and its "world-forming" powers, to take up the task of "speculating once more about the nature of reality independently of thought and of humanity more generally."[33] This realist position, of course, exemplifies the very definition of *dogmatism* in Kant, asking after the God's-eye view that virtually all continental philosophy since the nineteenth century has roundly rejected. If anticorrelationists in turn reject anything uniformly, it's the post-Kantian insistence that any question about the nature of reality has simultaneously to take account of the subject posing the question, a drama that Harman dubs the "dogma of human access."[34]

Whatever else one might say about it, this question of "access" remains—on the face of it at least—completely central to phenomenological and deconstructive discussions of "world" (and thereby it remains central to the definition of life itself). As we have seen in Heidegger, and likewise in Derrida, the question of "life" will find itself posed not in realist (or what Heidegger derisively calls "metaphysical") terms but in relational terms, which ask after any given life-form's access (or lack thereof) to futurity, possibility, or self-transformation.[35] No question of life without the simultaneous ("correlationist") question of the living thing's access to this enigma called life: so, the Heideggerian-Derridean question is not "What is life itself?" but "What's the relationship of any given form of life to the potentialities of world that define a living being?" Indeed, Jakob von Uexküll, author of the hugely influential *A Foray into the Worlds of Humans and Animals* (and a privileged source for Heidegger), offers us a crystalline example of correlationism in his methodological introduction to *Theoretical Biology*:

> No attempt to discover the reality behind the world of appearance, i.e. by neglecting the subject, has ever come to anything, because the subject plays the decisive role in constructing the world of appearance, and on the far side of that world there is no world at all. All reality is subjective appearance. This must constitute the great, fundamental admission even of biology. It is utterly vain to go seeking through the world for causes that are independent of the subject; we always come up against objects, which owe their construction to the subject.
>
> When we admit that objects are appearances that owe their construction to a subject, we tread on firm and ancient ground, especially prepared by Kant to bear

the edifice of the whole of natural science. Kant set the subject, man, over against objects, and discovered the fundamental principles according to which objects are built up by our mind. The task of biology consists in expanding in two directions the results of Kant's investigations: (1) by considering the part played by our body, and especially by our sense-organs and central nervous system, and (2) by studying the relations of other subjects (animals) to objects.[36]

There's much for a Derridean analysis to linger over here, but let's zero in on Uexküll's most fiercely correlationist claims about human subjects and their world: "the subject plays a decisive role in constructing the world of appearance, and on the far side of that world there is no world at all. All reality is subjective appearance." This represents the "hard" correlationist position, and I think it's safe to conclude that Heidegger, and with some caveats Derrida, is in agreement with Uexküll's first sentence here: the living subject has a special "constructing" relation to the "world of appearance" (Derrida's caveat, of course, being with the living *human* subject's exclusive access to world-formation); and it certainly seems that Derrida's thematization of death as "each time unique, the end of the world" would dovetail quite nicely with Uexküll's sense that "on the far side of that world there is no world at all." But the agreement ends there. Neither Derrida nor Heidegger would affirm where Uexküll immediately goes from there, to the monster correlationist claim par excellence: "All reality is subjective experience."

So how can we differentiate the first claim made by Uexküll from the second, insofar as Meillassoux's analysis would suggest that linkage between the two claims is ironclad, what you might call the correlationist two-step: world is an irreducibly relationist or correlationist concept, naming a living entity's openness or access to its own unique possibilities (which of course ends each time with death); hence the second step, which (for the anticorrelationist) seems to follow inexorably from the first: there is no reality worth the name but the reality of the individual living being (which Uexküll is also quick to admit could be an animal—so simply extending the concept of world to animals or other living things doesn't answer the charge of correlationism).

However, as I hope to have demonstrated through the analysis of *Glas*, at least one proleptic Derridean reply to the correlationist challenge is deceptively simple: for Derrida the question of life is not equated with the question of the real. Everything is subject to time, emergence, and the structure of the trace (there's your "real"), but not everything is alive (as I suggest, not even plants, technically speaking). Hence not everything has a correlate phantasmatic "world," and not everything "dies." Recall that *life*, through the crucially related term of *world*, is defined by a singular being's relation to time, futurity, and finitude. Indeed, it would seem that much of the vegetable kingdom fails the test of life not because those entities lack agency, sensation, communication, or whatever but because plants are not always singular beings. Only singular organisms can have phantasmatic worlds and therefore be alive.

In any case here's the punch line of all this: I think one can grant to Meillassoux that life is inexorably a correlationist discourse. However hard it is to define, the living is in practice most often differentiated from the nonliving precisely by the presence of a feedback loop of some kind (phantasmatic or otherwise); or at least virtually all definitions of life contain some defining sense of desire and response to immediate surroundings (to a thing's "world"): feeding, growing, reproducing, stayin' alive. In sum, as Thacker points out, "Every ontology of 'life' thinks of life in terms of something other-than-life."[37] Given that fact, I think it's easy to grant that life as we know it is correlationist (necessarily relational in terms of the thing that's alive), without having to accept anything of the second part of Uexküll's grand correlationist bargain: that reality is simply equitable with a subjective experience of it (human or otherwise). Arguing that life remains inexorably defined by multiple relationalities does not necessarily commit you to agreeing that "all reality is subjective experience." Life as we know it is certainly tied up with a singular entity's striving to persist in time; but it's certainly not the case (in Derrida, at least) that there is no reality without this particular sense of life. There exist myriad things that are not alive in Derrida, but that hardly means that they don't participate in reality.

The indifferent, violent emergence and passing away of *physis* or *différance* is perhaps a name for the real in Derrida; but *physis* is not alive, nor is *différance*. And in the end there's no "correlating" with *physis*, insofar as this radically neutral principle of emergence (this "superpower") is hardly beholden for its existence to human or animal life. Quite the opposite, as we have seen: *physis* predates (is in Meillassoux's terminology an "ancestor" to) this or that being, species, or form of life. And as I try to show above, *physis* is consistently yoked in Derrida to the question of emergence, especially in the text of *Glas*—which demonstrates in its own odd but inexorable way that nutritive, vegetable life is the condition of (im)possibility for the human and animal life of striving desire.

Of course, it's scientifically inaccurate to say that plants emerged on earth before land animals (because they almost certainly didn't); and even to argue that plants did come first in some way (if you count microscopic blooms in the

sea) doesn't solve much in a "what's before what?" argument about the evolution of animal and human life on Earth, insofar as the single-cell kingdoms
emerged hundreds of millions of years before anything else. Something was
always already before life, however you define it. Additionally, none of those
evolutionary discoveries about the emergence of life seems well positioned to
answer the correlationist "ancestor" argument put forward by Meillassoux—
that shit happened (presumably according to certain scientific principles that
we can now describe) for a few billion years on this planet before there was
any singular entity or living being around to take notice of it, and likewise for
nearly fourteen billion years in the rest of the universe. So, Meillassoux suggests, to insist that knowledge or science is a human construct (as defined in
the Kantian Copernican revolution, refined by Nietzsche as the simple premise
that knowledge is merely what we add to things) seems wrongheaded. Additionally, to say that plants are not alive, but are the condition for all life, seems
flat out absurd, from a scientific or realist point of view.

So, to clarify: plant life does not name the scientific kingdom Plantae for
Derrida, but plant life functions as an intense figure for this distributed, violent
power of emergence on which everything depends. The earth itself emerged,
after all. Recall again his discussion in *The Beast and the Sovereign* of "a phenomenon of *physis* like that young sprout or that (primarily vegetable) growth,
that partheno-genetic emergence we talked about when we were marking the
fact that, before allowing itself to be opposed as nature or natural or biological
life to its others." And plant life is certainly not the only figure in Derrida for
this overarching "superpower" of emergence and ruin (the condition of possibility that is simultaneously the condition of impossibility). There is a certain
way in which Derrida's entire project is offering a series of names for this X of
the real (*différance*, pharmakon, chora, the trace, the always already, the performative, auto-immunity, originary technicity, the event, and so on). Derrida
never ceases to offer us iterative names for these inescapably temporal conditions of emergence that guarantee only one thing: the impossibility of transcen-
dent, atemporal, "full" presence. Plants are a conceptual figure in Derrida for
the power of emergence, but they are not emergence "itself," as emergence is
not something that can be captured, by definition: such emergence is each time
unique, the beginning of what may or may not become a world.

But, at least for me, part of the stake of staging and exploring this plant-
versus-animal face-off in *Glas* is an attempt, to use the current lingo, to question the correlationism that seems inherent in figuring the animal as the

shattering other of the human. Animality remains a biocentric concept, or at least it does in Derrida, and it thereby remains dependent on a kind of correlationist relation to world, because the discourse on life presupposes that being or possibility or futurity or even death (a thing's "world") is manifest within (and directed at) a singular "living" entity of some kind. If it's alive, it has a world, and vice versa. This is axiomatic, or so I'm claiming, in Derrida. But of course "life" (in its correlationist relation to world) isn't all there is in Derrida; life is not a master term in his thinking.

Vegetable life is not the only name for emergence itself (any more than *différance*, the trace, or any of the other names Derrida offers), but plant life remains an intense and provocative instance of emergence. And such vegetable *physis* is compelling among those other names precisely insofar as plant life is familiar to us, ordinary in the best sense; but vegetable life remains blocked in interesting and productive ways from an easy analogy or correlation to the human and/or animal life trajectory of world and relation.

Life in Derrida, in the end, is exceedingly rare, hardly the "everything" of correlationism. This makes Derrida's view of life less correlationist, vitalist, or realist than, as Henry Staten suggests, "naturalist"—materialist with a twist (knowledge with a tinge of faith, one might say, following Derrida's great "Faith and Knowledge" essay). As Staten writes, "the strong naturalist view, from which Derrida does not deviate, holds that matter organized in the right way brings forth life, but denies that life is somehow hidden in matter and just waiting to manifest itself. . . . Life is a possibility of materiality, but as a vastly improbable possibility, by far the exception rather than the rule."[38] Staten here provides a convincing reading of Derrida on life as a "vastly improbable possibility," but this sense of life does (I will point out in closing this chapter) completely reinscribe the very murky je ne sais quoi—the transcendental horizon—that Foucault diagnoses as the discursive move that birthed "life" as we know it in the nineteenth century. Life in Derrida remains defined as a singular finite entity's relation to being and to a realm of world-as-possibility. And for Derrida life remains, as Staten points out, a rare nearly mystical occurrence, a "vastly improbable possibility."

I should also note in closing this chapter that the conditions of possibility for life (a relation to world) seem taken for granted in much biopolitical discourse—suggesting that world or possibility or futurity is secured, and that our nearly miraculous relation to that phantasmatic world is the transient factor in the game of life. The condition for the possibility of life is not alive, but it remains secure, maybe even transhistorical. It seems increasingly clear, however, that "world" (as in our world, this world) can "die." The earth could be (eventually will be) reduced to an ash-heap, either by our own hands or by (inter)planetary changes of some unforeseeable kind. And if world can die, that would seem to confirm world itself is in some way "alive."

Likewise, this analysis would seem to deepen our Foucauldian suspicions about contemporary philosophy's refusal or inability to think historically about "life" (the stubborn sense that life is now, always was, and always will be primarily about a desiring relation to world—life everywhere entails a singular being that cares about maintaining itself across a desiring interval). As I've argued here, you can't really think about the vegetable life of ecosystems on these terms—again, less because of the "striving" requirement for life than because of the "singular entity" requirement. In the end Derrida's rich and provocative work on life and world has the ongoing effect of continuing to contain the question of life within supposedly singular organisms and their worlds (phantasms though they both clearly are in Derrida's work). As we will see when we turn in the next chapter to Deleuze and Guattari, these notions of phantasmatic world and individual organism increasingly seem neither necessary nor sufficient conditions for thinking about life in the twenty-first century.

4

FROM THE WORLD TO THE TERRITORY

Vegetable Life in Deleuze and Guattari; or, What Is a Rhizome?

ASKING "WHAT IS A RHIZOME?" in Deleuze and Guattari may seem like a flat-footed, wrongheaded, or merely superfluous question for one or several reasons. First, the rhizome in Deleuze and Guattari covers territory that's been very thoroughly commented on, maybe even buried under scholarship, in the past couple of decades. So, really, what's left to say about the rhizome (or, one might wonder, about any of the foundational Deleuzoguattarean concepts)? Second, the Aristotelian form of the question itself ("what is x?") may seem hopelessly old-fashioned, despite Deleuze and Guattari's attraction to this very phrasing—in their last work, for example, *What Is Philosophy?* Besides, as I noted with the first objection, we already know full well what the rhizome is: multiple, intense, subterranean, resistant, connective, smooth, molecular, a block of becoming. (And thereby we also know what it's not: totalized, extensive, arborescent, compliant, closed, striated, molar, a block of being.) What is a rhizome? Quite simply, it can seem like the rhizome is one-stop shopping for all your deterritorialization needs. As Elaine Miller writes in *Vegetative Soul*, "Rhizomes have no territory; they may spread with the wind and cover any amount of ground, since they grow outward rather than upward."[1]

Against this grain I'm going to stress here that the rhizome is equally (and in fact more importantly) a matter of territorialization, mapping a way of inhabiting territory—of "living"—that's crucially different from the fictional phantasms that Derrida sees as necessary to any living being's "world." Rhizomatics constitutes a style of conceptual personae, in Deleuze and Guattari's sense: a field of forces that dramatizes a new image of thought and concept of life, outside the organism-centered, human-animal understandings of life that continue to dominate our biopolitical present.[2]

And there's perhaps an even more awkward consequence to raising the question "What is a rhizome?" at this point in our investigation: the rhizome seems the obvious place to *begin*, rather than *end*, a book concerned with plant life in Foucault, Derrida, and Deleuze and Guattari. I hope I've managed to shed some obscure light on the myriad roles of vegetable life in selected texts written by Foucault, Derrida, and some of their interlocutors (Agamben and Heidegger, among others). But surely it's not going to be difficult to demonstrate a foundational debt to plants in Deleuze and Guattari's work: the rhizomatic life of plants is nothing less than the opening gambit and the virtual map to their most famous work, *A Thousand Plateaus*. "Introduction: Rhizome" is the plateau that you read first and is in fact the chapter that says, when you're done with it, you can read the rest of the book in any order you like.[3]

In short, Deleuze and Guattari's engagement with plant life does not, unlike Foucault's or Derrida's, seem to require much unearthing or unpacking. The centrality of vegetable life hardly remains obscure or hidden in a discourse that urges us straightforwardly to "Follow the plants" (*TP* 11), even though "it's not easy to see the grass in things" (*TP* 23). Arguing that the rhizome (and by extension, plant life itself) remains important to Deleuze and Guattari can then seem a bit like peddling water at the mouth of a mountain spring: a hard sell precisely because it's so obvious and ubiquitous. Deleuze himself characterizes "the research undertaken with Guattari" straightforwardly as "a vegetal model of thought."[4]

But I'd like to suggest that this consensus concerning vegetable life's centrality in D&G (however undoubtedly "correct" it is) constitutes not so much a solution as a problem for thinking through the roles of plant life in their work. The primary difficulty, I think, is this: plant-based rhizomatics has in the present scholarly context become a metaphor for everything and anything—or, more precisely, a metaphor for how everything is connected in an underground way, linked to a font of hidden, living fecundity just below the surface (doubling down on that "first birth of biopower" that we encountered by way of Foucault earlier in this book). So, for example, we've heard that the Internet is a rhizome, as is sexuality, city planning, both Jack Kerouac's and Dr. Seuss's writing, education, disability studies, Facebook and Twitter, robots, action research (whatever that is), the movements of aboriginal peoples, poetry, performance art, community service learning, wikis, nursing, sustainability, fitness, and so on. (Just do what I did: run "Deleuze" and "rhizome" through Google—itself dubbed "rhizomatic" in at least one article—and see for yourself.) Ironically enough, rhizomatics has become a template for discussing virtually everything *except* plant life.[5]

So with your indulgence, I'd like to revisit the question of rhizomatics, for the time being naively bracketing the myriad analogical connections already forged in the scholarly discourse surrounding the topic. Aside from my own narrowcast argumentative purposes here, there likewise exists a warrant within Deleuze and Guattari for skepticism concerning this runaway metaphorics surrounding plants and rhizomes. To put it bluntly, when you say "X is like a rhizome" (which is what I presume most people mean when they say that Facebook or performance art is "rhizomatic"—while often quite dull, checking Facebook or watching performance art is seldom as literally excruciating as watching a plant grow), it's the metaphorical "like" that constitutes the major Deleuzoguattarean problem: throughout the breadth of their coauthored work, they have nothing (but nothing!) nice to say about metaphor. In *A Thousand Plateaus*, for example, they speak of "the abolition of all metaphor" (*TP* 69) and say quite unequivocally that "if we interpret the word 'like' as a metaphor or propose a structural analogy of relations . . . we understand nothing of becoming" (*TP* 274).

Becoming isn't well understood metaphorically primarily because becoming in Deleuze and Guattari is not a "freeing" motion (a fluid state of deterritorialization opposed to a state of striated, territorialized being) but, oddly, the other way around: which is to say, becoming is primarily a process of inventing and inhabiting territory otherwise (rather than a preexisting thing freeing itself from its present stratification by becoming "like" something else). Deleuze and Guattari borrow their notion of becoming primarily from Gilbert Simondon, who describes it like this:

> This division of being into phases is becoming. Becoming is not a framework in which being exists, it is a dimension of being, a mode of resolution of an initial incompatibility that is rich in potentials. *Individuation corresponds to the appearance of phases in being that are the phases of being.* It is not a consequence placed at the edge of becoming and isolated; it is this operation itself in the process of accomplishing itself. It can only be understood on the basis of the initial supersaturation of being—without becoming and homogeneous—that then structures itself and becomes, bringing forth individuation and environment, according

to becoming, which is a resolution of the initial tensions and a conservation of these tensions in the form of structure.[6]

In short, becoming is not primarily a matter of freeing an organism from the grip of a limiting determination but is better characterized as a process of "individuation"—a "mode of resolution" within a field of being rather than the undermining or outflanking of reification. Becoming, in short, is not opposed to being (as, for example, an open system is to a closed one or as the sensible is opposed to the intelligible in metaphor); rather, becoming is the primary modality of the being's segmentation, the process whereby this stuff called being (which in this thematization is largely free from determination—"pure immanence" is what Deleuze and Guattari call it) "structures itself and becomes" as an emergent series of individuations.

Becoming then is constituted as a series of striations or phase transitions within a constantly singularizing entity that's understood less as a preexisting *form* (or organism, if it's a living thing) than as an ongoing series of *variations*— a subject, a crystal, a plant, or a hurricane. In Deleuze and Guattari's account such individuations emerge and become through a self-organizing patterning of micrological harmonizations—what D&G famously call refrains—rather than their becoming having the quality of freeing of preexisting entities from the iron cage of their totalizing form (e.g., "the human" freed by "becoming-animal"). Such individuated becoming, in other words, is not an outside that an already-configured singular entity or organism strives toward, but it is the mode in which anything like an "entity" arises and transforms in the first place. For Simondon, as for Deleuze and Guattari, life isn't trapped inside static forms but is characterized by "perpetual individuation, which is life itself."[7]

Just as important for our "worldly" purposes here (coming out of the Derrida chapter), note that such individuation or becoming doesn't primarily emerge within (nor in relation to) a phantasmatic *world*. Rather, any individuation coemerges with what Deleuze and Guattari call its *territory*: becoming is a process of "bringing forth individuation *and environment*." An entity's "world," in other words, is not a necessary, enabling backdrop against which "shit happens," nor is it a phantasmatic projection or enabling fictional field of possibilities that living things invent for themselves. In the process of becoming, the world (remapped onto the terrain of territory) coappears with individuation rather than serving as the opening of a field where such an individual emergence might occur. Simondon elaborates: "The very notion of a qualitative or intensive series should be thought according to the theory of the phases of being:

it is *not relational* and is not maintained by a preexistence of extreme terms. . . . One must begin with individuation, with being grasped at its center according to spatiality and becoming, not with an *individual* that is substantialized in front of a *world* that is foreign to it."[8] It is through consolidation or harmonization of refrains—what Deleuze in *Difference and Repetition* will (somewhat oddly) call "contemplation" or the extraction of "habits"—that entities come into being and continue to transform. And within such a regime neither the entities nor their territories for living (neither the *substantive* nor its *world*) preexist their coemergence. A tornado doesn't grow *out of* a thunderstorm but is constituted by an intensification of forces *within* one—a phase of transition that brings forth both a new territory and a new entity called the tornado. The interaction of the orchid and the wasp famously gives rise to an individuated territory that is no longer orchid nor wasp. John Protevi sums up the stake of this sense of becoming: "For Deleuze, both realism and idealism are flawed because they take one side of an opposition of finished products: a fully formed world or a fully formed subject."[9] So, unlike the project of deconstruction, setting out to undermine the reification of "fully formed subjects" is hardly the prey of Deleuze and Guattari's work; rhizomatics is not useful or interesting because it's a modality characterized by liberation from determination but precisely because it's a novel modality of continuous individuation or striation of a territory for living.

So my query "what is a rhizome?" might profitably be rephrased as, "what is a rhizome if it's not primarily a *metaphor* for x, y, or z 'liberating' formation, practice, or thing"?[10] Here I'd like to contest (or at least nuance) this dominant sense of rhizomatics, summed up by Laura Marks when she argues that "Deleuze and Guattari invite humans to be weed-like and celebrate 'drunkenness as a triumphant irruption of the plant in us,' an interconnected receptivity. We would like to become more plant-like,"[11] or when Marder writes that for D&G "to live is to be superficial and dis-organized: to exist outside the totality of an organism: to be a plant."[12]

Concomitantly, we might ask, what is plant life in Deleuze and Guattari if it's not merely a utopian linguistic figure for something we should strive to become—a more open, free, "better" version of posthumanist subjectivity? What is becoming if it's not a phantasmatic Derridean "world" of "disorganized" possibilities projected and sought after by living things?

Importantly enough, the concept of the rhizome is introduced in the opening pages of *A Thousand Plateaus* precisely as a counterpoint to the overarching metaphorical model of the tree that's dominant in the human sciences, most prominently in biology (with its roots and branches of life), philosophy (truth trees), and linguistics (the trees of language development, from the roots of proto-Indo-European through the trunks of Greek and Latin to the branches of the modern European languages). Recall that *A Thousand Plateaus* opens with Deleuze and Guattari very specifically introducing the rhizome in order to limn out two versions of a book: a root-book and a rhizome book, the work understood on the model of a tree versus the text as a rhizome. As they insist in these opening pages, "thought is not arborescent": "we're tired of trees" (*TP* 15), they infamously lament. This opening distinction and bifurcation of plant life (rhizomes pitted against trees) has led some very smart commentators—among them Steven Shaviro, David Wood, and Michael Marder[13]—to wonder how D&G could be so dismissive of trees, which are of course plants themselves and thereby offer a kind of asubjective antidote to traditional human-animal limits of philosophical thinking that the rhizome is presumably hatched to subvert (not to mention the botanical fact that some trees, like aspens for example, are also rhizomatic, replicating along a line of stolon growth rather than reproducing by seed distribution). As Wood wonders about their discussion of trees and rhizomes, "this binary opposition sounds all too like the tree logic that Deleuze and Guattari are arguing against. . . . If Deleuze and Guattari are right about trees, then trees are not trees. And to borrow Hegel's voice here for a moment, we might say that the rhizome is the truth of the tree."[14]

So, following Shaviro, Marder, and Wood (and perhaps building on work in animal studies that defends companion animals against D&G's accusation that "anyone who likes dogs or cats is a fool" [*TP* 240]), one could mount a vigorous defense of trees here—hoisting the protest sign, "D&G Unfair to Dogwoods." But I'm not going to do that, mostly because I don't think this rhizome/tree distinction is precisely to the point in thinking about vegetable life in D&G. Rather, I think we need to revisit their investment in vegetable life and the rhizome specifically in its context of its singular emergence within *A Thousand Plateaus*, where the rhizome functions as (among many other things) another weapon in their career-long battle against metaphor, against interpretation, against a conception of world as a series of hidden possibilities. In short, their issue with the tree as an "image of the book" is roughly the same as the problem with Ernest Hemingway's analogy of the iceberg as an image of fiction: while it sounds good (who doesn't love stuff that's "deep," mostly hidden below the surface?), in practice it consistently encourages swapping out the surface movements and connections of a text (following the surface connections of a territory along a line of flight) for an excavation operation, where all the action is conveniently hidden in a misty realm that requires hermeneutic interpretation and the projection of a world.

At the outset of *A Thousand Plateaus* D&G insist that, like everything else, books are not fictional worlds but territories made up of roots and rhizomes, layers of strata and lines of destratification: "In a book, as in all things, there are lines of articulation or segmentarity, strata and territories; but also lines of flight, movements of deterritorialization and destratification. . . . All this, lines and measurable speeds, constitutes an *assemblage*. A book is an assemblage of this kind" (*TP* 3–4). In this opening context (a metadiscussion of what kind of book *A Thousand Plateaus* will be) their critique of trees is most immediately connected to their ongoing critique of meaning, signification, and lack: "We will never ask what a book means, as signified or signifier; we will not look for anything to understand in it. We will ask what it functions with, in connection with what other things it does or does not transmit intensities. . . . A book itself is a little machine . . . Writing has nothing to do with signifying. It has to do with surveying, mapping, even realms that are yet to come" (*TP* 4–5). So this is the specific context in which all this "root-tree vs. rhizome" controversy emerges in D&G, within a discussion of the hidden organicism of many reading practices: "A first type of book is the root-book" (*TP* 5), which is critiqued by D&G because of its shopworn, monotheistic nature: "one becomes two" (*TP* 5). From the book's seed, from the unseen origin of hidden meaning, something large and imposing can be projected (an entire world of commentary and interpretation, for example), where the visible, affective or effective part of the organism-book remains tied to (and can ultimately be explained by) its unseen origin (roots, seeds): "In order to arrive at two following a spiritual method it must assume a strong principal unity" (*TP* 5).

The second type of root-book is somewhat more molecular, emphasizing a more diverse fascicular or radical system akin to grafting—which maps a reading or writing practice wherein a root transforms when a "secondary root grafts onto it and undergoes a flourishing development" (*TP* 5). But in the end this seemingly more fragmented or multiple radical system book remains in practice a root-book. And here D&G thematize the breakage of the original root, the frustration of the move to the origin, as a kind of double-down on the maneuver of protecting hidden depth in the service of projecting an open world: "This is why the most resolutely fragmented work can also be presented as the Total Work or Magnum Opus. . . . The world has become chaos, but the book remains the image of the world: a radical-cosmos rather than a root-cosmos. A strange mystification: a book all the more total for being fragmented" (*TP* 6).

As an example of this fragmentation-as-higher-unity, Deleuze and Guattari cite the examples of Friedrich Nietzsche's aphorisms, James Joyce's *Ulysses* and *Finnegans Wake*, and William Burroughs's cut-ups; but maybe the clearest indication of how fragmentation itself can become a new unifying principle is TS Eliot's "The Waste Land." With these "fragments I have shored against my ruins," Eliot's modernist masterpiece suggests that the only way "out" of the fragmentation portrayed in the poem is the higher, discontinuous unity of the poem itself, a modernist aesthetic practice that can transform "a heap of broken images" into a world seething with possibilities. Like other modernist works (think of Van Gogh's great serial paintings of *Peasant Shoes* or the dice-throw in Mallarmé, all the way to Beckett's plays), it is only the possibility-laden world of the artwork itself that can offer hope for redeeming, transforming, or simply allowing us to live through the broken, melancholy world depicted in the artwork. In any case if I insist on revisiting the singular emergence and functioning of rhizomatics within D&G's texts (the rhizome's individuation with the territory of *A Thousand Plateaus*), it's at least partially to suggest that quite a bit of the secondary literature falls under their diagnosis of the high modernist, "radicle" category of the root-book, continuing to understand rhizomatics as a kind of fragmentation-as-higher-unity, as the projection of another world. For D&G, art constitutes a terrain for living, not primarily a phantasmatic representation of hidden or forgotten possibility.

The stake of the distinction between tree and rhizome is less a matter of bifurcating plant life into two projected worlds—the deterritorializing rhizome (good) and the striating tree (bad)—and more a matter of understanding rhizomatic plant life as a kind of "inorganic" life characterized less by projection of a world than by the incessant becoming of individuation, the coemergence of the entity with its territory. This is to say, rhizomatic plants in D&G function not as organisms or things with a world (that's the tree) but as a machinic power of emergence within a territory, as a robust "inorganic" sense of life. And I would emphasize that *inorganic* here is not a metaphor (for machinic excess or the raw materiality of vibrant matter itself) but functions quite literally. Simply put, "inorganic life" indexes a notion of life (and its territory) not housed exclusively in or by organisms (and their worlds).

To fold this analysis back onto the secondary literature, Deleuze and Guattari's description of the root-book (and its projected world of possibilities) also nicely encapsulates much of what the secondary literature does with rhizom-atics in D&G—and likewise shows us that plants can function perfectly well as poster children for an ironically organicist posthumanism, just as Foucault suggests that animals have. If, for example, as Kari Weil argues in *Thinking Animals: Why Animal Studies Now?*, animal studies exists to "enhance the lives of all animals, including ourselves"—if animal studies' charge is indeed "to further humanism's noble aims"[15] of liberation for all beings—then there's no reason why plants can't function just as well as animals have as the "noble," revolutionary, hidden, forgotten or excluded other to the human—as repre-sentative of a better world for humans to inhabit. As Marder phrases this senti-ment, "the suggestion that the plant is 'a collective being' implies that its body is a non-totalizing assemblage of multiplicities, an inherently political space of conviviality."[16]

But the rhizome is introduced in *A Thousand Plateaus* not as a figure for a hidden protean excess that transgresses all normative categories, nor as a pic-ture of what the projected human world could become if we were more attuned to our others, much less as a hidden subterranean multiplicity-as-unity (those are the characteristics of the fascicular or second type of root-book, after all), but through a formula: the infamous "n–1." The multiple, Deleuze and Guattari insist, can't be incanted into existence (with clever slogans and typography): "the multiple must be made, not always by adding a higher dimension, but rather in the simplest of ways, by dint of sobriety, with the number of dimen-sions that one already has available—always n–1 (the only way the one belongs to the multiple, always subtracted). Subtract the unique from the multiplic-ity to be constituted: write at n–1 dimensions. A system of this kind could be called a rhizome" (*TP* 6). While it functions as a kind of definition, this math-ematical formula (n–1 = rhizome) hardly seems to clear up all the ambiguities surrounding the question, "what is a rhizome?" So they continue: "A rhizome as a subterranean system is absolutely different from roots and radicles. Bulbs and tubers are rhizomes. Plants with roots or radicles may be rhizomorphic in other respects altogether: the question is whether plant life in its specificity is not entirely rhizomatic. Even some animals are, in their pack form. Rats are rhizomes" (*TP* 7). Well aware of the obscurity of this supposed clarification (all plant life—including trees—is probably rhizomatic, and even some animals are?), they go on to admit that "we will convince no one unless we enumer-ate certain approximate characteristics of the rhizome" (*TP* 7). So D&G begin again limning out the contours of the rhizome, but they remain faithful to their opening critique of hidden depth and metaphor throughout. One might say in fact that it's the metaphorical hiddenness of meaning—the sublating, asymmetrical unity of the organism in relation to its world—that must be subtracted in their method.

In fact, n−1 might be narrativized like this: subtract the hidden, unifying power of the organism (and/or its sibling, the projected world) from any given entity *n* to find the rhizome lines of individuated emergence. As Gregg Lambert writes, it's a matter of "subtracting this final image or higher dimension of unity or totality: transcendence."[17] Or as Deleuze phrases it concisely in *Difference and Repetition*: "Every time we find ourselves confronted or bound by a limitation or an opposition, we should ask what such a situation presupposes. It presupposes a swarm of differences" (50). For D&G the tree remains a problematic diagram or grid precisely because it's a vertical picture of a metonymic linkage; the tree is a metaphorical world—a noun, an organism, the sublation of a "swarm of differences" rather than the continuing variation that constitutes such a swarm of processes. In fact, the tree is a metaphor for metaphor (the movement from the sensible to the intelligible): the insistence that everything living is housed within organisms and worlds, and is therefore traceable through descent to a hidden, protean root/origin. What's under pressure in D&G's rhizomatics is a notion of life as housed entirely within organisms, which function as the *n* of "n−1":

> Any point of a rhizome can be connected to any other, and must be. This is very different from a root or tree, which plots a point, fixes an order. The linguistic tree on the Chomsky model still begins at point S and proceeds by dichotomy. On the contrary, not every trait in a rhizome is necessarily linked to a linguistic feature: semiotic chains of every nature are connected to very diverse modes of coding (biological, political, economic, etc.) that bring into play not only different regimes of signs but also states of things of differing status. . . . A rhizome ceaselessly establishes connections. (*TP* 7)

The initial (though by no means exclusive or final) stake of the rhizome/tree discussion in *A Thousand Plateaus* concerns language—the stuff of the book—and the ways in which it moves and reproduces rhizomatically, not through a seed and root structure like a tree. Like any rhizomatic entity, language is a machinic and connective assemblage of disparate microprocesses, therefore "alive," but in an inorganic sense. Like a rock or a river, language is a connection machine that creates simultaneous territories and individuations; language is not merely or primarily a discrete entity with a protean core of excessive life (multiple meaning) fighting against its imprisonment by this or that constricting interpretation. As they write, in terminology recalling *Anti-Oedipus*'s connective synthesis, "the tree imposes the verb 'to be,' but the fabric of the rhizome is the conjunction, 'and . . . and . . . and . . .'" (*TP* 25). It is this machinic "and . . . and . . . and . . ." that constitutes the process that goes by the name "a life" in D&G, and such a swarm of singular processes that "makes" any given individual and its territory.

In short, n−1 limns the ways that an individuation (coemerging with a territory) subtracts a repetition from difference, a habit from the chaos of being, a life from "life." They clarify this strange machinic vitalism further on in *A Thousand Plateaus*: "If everything is alive, it is not because everything is organic or organized but, on the contrary, because the organism is a diversion of life. In short, the life in question is inorganic, germinal, and intensive, a powerful life without organs, a body that is all the more alive for having no organs, everything that passes between organisms" (*TP* 499). If nothing else, this suggests a Deleuzoguattarean notion of "life" that's radically distributed in a field of individuations, not "fed" by roots, origins, or the unfathomable depths of a protean, excessive possibility out of which new entities find themselves to be replicates birthed by preexisting organisms: "Multiplicities are rhizomatic, and expose arborescent pseudomultiplicities for what they are. There is no unity to serve as a pivot in the object or to divide in the subject" (*TP* 8). There is, in other words, no preexisting living organism to serve as the metaphorical wellspring of life—as the ancestor who gives rise to its progeny through reproduction and mutation. Rather, striations of preexisting lines of flight are what we come to call organisms, species, things—and there's nothing "wrong" with that, aside from the sense that such hypostasization tends to obscure the nature of individuation and life (which is an ongoing series of microprocesses). In *Difference and Repetition* Deleuze sums up the consequences of this position for human subjects quite simply: "The self does not undergo modifications; it is itself a modification" (79). We are a smear of emergent singularities rather than a group of evolutionary continuations or mutations of an existing ancestral line.

While the rhizome in D&G is not to be confused with life itself, their insistence on the "inorganic" nature of rhizomatic multiplicity is what connects all this talk about inorganic life to Foucault's work on the birth of biopower: both Foucault and Deleuze and Guattari are involved in a full frontal attack on a discourse of life as that murky, projected world animating any given "n"–a thing, relation, or being. "Life" moves and advances in D&G through the singularities of filiation, not through the hidden software of mass evolutionary descent—which is why, they consistently argue, "evolutionary schemas may be forced to abandon the old model of the tree and descent" (*TP* 10). All evolution is coevolution, and coevolution is rhizomatic.

At the end of the day it would seem that the rhizome "is" no more vegetable life itself than the rhizome "is" the Internet, Kerouac's novels, capitalism, or whatever—precisely because life in D&G names a swarm of inhabiting processes operating at a molecular level (not as a relation between an entity and its projected world). As an explanatory transversal connection on this point we could note that much of the continuing critique concerning D&G's "becoming-animal" revolves around a similar problematic: the questionable "organic" reading goes something like this: insofar as becoming-animal is becoming *an* animal or *like* an animal, then rhizomatics would or could entail becoming (like) *a* plant. As Keith Ansell-Pearson complicates that kind of understanding, he argues that in D&G "'animal becomings' . . . are to be conceived as taking place on a molecular level having little to do with the animal on the level of genus and species."[18] If this is indeed the case—that becoming is a molecular series of individuation processes, not an organic phenomenon of mutation manifest through reproduction or replication—then it seems that the whole metaphorical imitation problem (D&G ask us to become like a wolf or a bear or a daisy, to transgress the human form in order to fulfill our multiple animal or vegetable potentials) has been a red herring from the beginning.

When D&G talk of becoming, or when they talk of animals or plants as rhizomatic, they are referring not so much to this or that privileged or exemplary organism (the leaves of grass or the rats, the orchid or the wasp) as they are conjuring the multiple processes (the territories of becoming) that traverse any given phenomenon or organism—the swarms of white blood cells in mammals, the global winds that blow the dust from the Sahara to the rain forests of South America, the lightning that causes brush fires, the abundance of prairie life that follows a decade after such fires, the interplanetary forces of gravity, or the odd brain chemical effects coming from the ingestion of THC or poppies. These rhizomatic becomings are nonorganic bases of the organism—the swarm of verbs that literally make up the nouns of organisms, subjects, or objects: "a machinic network of finite automata (a rhizome)" (*TP* 18).

As Ansell-Pearson outlines the stakes of redefining life in this molecular manner, the analytic driver for their project is not an evolutionary striving to-

FROM THE WORLD TO THE TERRITORY

ward a beyond or some barred-off "outside" world of heightened awareness or affectivity—which is a popular way of thinking about "becoming-animal," or the sense that Laura Marks conjures when she suggests that in D&G's rhizomatics "vegetable locomotion critiques normativity, at least when plants are left to their own devices. Plant life invades the privilege normally accorded to animals. Weeds breach norms of human culture."[19] Such a suggestion of transgression toward an unattainable outside (in the direction of a protean and unbounded, more meaningful or higher "world" outside the organism) is largely rejected by D&G. Their focus goes in the opposite direction, characterized by an involution toward the microprocesses where multiplicity is less a murky, unthematizable "beyond" to be striven after than it is a swarm of assemblages already operating within substantives in a concrete way. As Ansell-Pearson argues, "the abyss of the Dionysian world . . . is not, Deleuze insists, an impersonal or abstract Universal *beyond* individuation. . . . It is rather the 'I' and the self that are abstract universals that need to be conceived in relation to the individuating forces that consume them."[20] This process of involution was of course what psychoanalysis had right all along—"the rhizome is precisely this production of the unconscious" (*TP* 18)—though unfortunately all psychoanalysis ever found at the level of the unconscious was more trees: "On the verge of discovering a rhizome, Freud always returns to mere roots" (*TP* 27).

Deleuze phrases this somewhat esoteric point quite straightforwardly in *Difference and Repetition*: "We are made of contracted water, earth, light and air—not merely prior to the recognition or representation of these, but prior to their being sensed. Every organism, in its receptive and perceptual elements, *but also in its viscera*, is a sum of contractions, of retentions and expectations" (73). As we saw in our examination of Aristotle's *De anima*, these swarms of molecular emergence have functioned as one of the privileged names of "vegetable life" throughout the history of Western philosophy, so it's not surprising that D&G dub them rhizomatic. The "vegetable soul" functions in Aristotle and Plato as the proper name for intense growth without entelechy, these multiple but basic processes of transformation and decay that are shared among all living things (else, on the standard definition, they would not be living). D&G take and then far extend this traditional portrait of vegetable life (as the most widely distributed form or definition of the living), rethematizing life as aggressively inorganic (not contained "in" organisms) and smearing it across all things. One can easily dismiss this position as "naive vitalism" (as many critics do), but D&G remain fine with that charge, at least understood on their own idiosyncratic terms: "Everything I've written is vitalistic," Deleuze insists in an interview; "at least I hope it is."[21]

Though as one has to repeat for the thousandth time, the rhizome/tree (smooth/striated, inorganic/organic) distinction is not merely a subset of myriad good/bad distinctions (with smooth rhizomes as the portrait of freedom, striated trees the iron cage of normative capture): "Of course, smooth spaces are not in themselves liberatory. But the struggle is changed or displaced in them, and life reconstitutes its stakes, confronts new obstacles, invents new paces, switches adversaries. Never believe smooth space will suffice to save us" (*TP* 500). D&G insist that what they call "anorganic life" is, in its most basic form, characterized by "chaos, pure and simple" (*TP* 503)—which is why stratification is not an evil to be avoided, but a necessity. As Claire Colebrook and John Protevi both maintain, in D&G it's always a matter of "a" life (the emergence of a singularity or becoming and thereby some striation), not so much a matter of life "itself."[22] Things coemerge with their striated territory; as such, living things or entities inhabit a territory more than they require a projected world.

So on the heels of our earlier discussion of Derrida, this is precisely why the question of "world" remains important to our reconsideration of life and biopower: for D&G the world is better understood as the entity's cocreation of a territory for living, the subtraction of a territory (a "difference"), or "contraction" of elements. In agreement with Derrida, one might say that in D&G a being and its world are coemergent, insofar as they are "each time unique," though for D&G this question of world is clearly not (as we saw above in Derrida) a phantasmatic projection reserved as the purview of living (which is to say, always-already-dying) organisms. Riffing on Uexküll's work concerning animal worlds during his *L'Abécédaire* interviews with Claire Parnet, Deleuze points out how "in a nature teeming with life, [the tick] extracts three things" to create its world: light, smell, and touch. "That's what constitutes a world," he replies simply.[23] If a world is then a *territory* for living, and life is pure immanence, then the world in D&G functions far differently than it does in Derrida (who, you'll recall, thematizes world as a mediating map that will forever delude living beings into thinking that it is the territory). In D&G the world is a territory that coemerges with individuated entities rather than a "false" but enabling fictional projection (available to some organisms and not to others).

For D&G the living organism doesn't have any particularly privileged access to the chain of territories, precisely because the question of territory is one of *inhabitation* rather than *access* (which remains the primary terrain of world).

And if for D&G any individuated being is coexistent or coemergent with its world, its territory, it may then follow that death in D&G is characterized not by the Derridean "each time unique, the end of the world" (the end any given organism's access to possibility) but rather as "each time unique, the end of the territory." As Protevi puts it, "Every iteration of a process, each case in a series of organic syntheses . . . has its own rhythms that allow cells to live. Death, we can speculate, occurs when the rhythms of the processes no longer mesh."[24] The territories or assemblages that allow individual entities to emerge and live are outcomes or orchestrations of microprocesses; when these orchestrations no longer "work," when the rhythms of passive synthesis can no longer mesh into a coherent territory for living, the entity itself ceases to exist.

One might say, in other words, that for D&G individuation is not a *condition of possibility* for any conception of life or world but a *condition of emergence* for this or that being, insofar as assemblages work to extract territory alongside individuation. So, for example, when Deleuze insists that three factors constitute the tick's entire "world," he suggests *not* that the entity "tick" projects this world (as a futural field of possibility) but that such a world is an assemblage of contractions or extracts from the chaotic mesh that is being: "assemblages are already different from strata. They are produced in the strata, but operate in zones where milieus become decoded: they begin by extracting a *territory* from the milieus. The first concrete rule for assemblages is to discover what territoriality they envelop, for there always is one: in their trashcan or on their bench, Beckett's characters stake out a territory" (*TP* 504). If world as a territory is made up of refrains and shot through with deterritorializing lines of flight, then the question of the world and its makeup is always a diagnostic one: "the concrete rules of assemblage thus operate along these two axes: on the one hand what is the territoriality of the assemblage? . . . On the other hand, what are the cutting edges of deterritorialization, and what abstract machines do they effectuate?" (*TP* 505): "rhizome lines oscillate between tree lines that segment and even stratify them, and lines of flight or rupture that carry them away" (*TP* 506). In short, D&G conceive of world as an assemblage made up of the individuated being linked to its territory, which is made up of lines of flight. Concomitantly, the world is not as a place where existing organisms can confront their ownmost phantasmatic (im)possibilities and constantly worry over their immanent death. To put it plainly, living entities in D&G are defined by what they *can* do (x, y, and z within their territory), not by what they *can't* do (survive forever, or gain access to possibility "as such"). Accordingly, the plant constitutes a strikingly different terrain from the life-territory of the (animal) organism. As Elaine Miller explains, "The plant is radically opposed to the figure of the organism as autonomous and oppositional; its stance toward the world is characterized by the promise of life and growth, not the avoidance of death or loss."[25]

While rhizomatics intensely traverses and characterizes the world of plant life (in the way, for example, that panopticism inhabits the prison intensely in Foucault), at the end of the day the rhizome *is* no more a blade of grass than being *is* Dasein in Heidegger, or subjectivity *is* a prison in Foucault. The rhizomatic relation remains on display very intensely in plant life, but vegetable life is not the only place where the rhizome abides. So, the logic of the rhizome is, perhaps, akin to Foucault's sense of the way that power traverses the panopticon: it's not that the prison is the secret metaphor for where we all live (as in D&G, it's not that the rhizome represents the hidden truth of all arborescent formations). And this obtains not so much because such formations about hidden truth are "wrong" but because they miss the stakes of the distinctions themselves. Just as in Foucault surveillance is the machine or practice that animates and saturates all the others (the panopticon is not a metaphor nor a phantasmatic dream building—it's an *operating system*), so in D&G rhizomatics names those swarming practices that saturate all the others—as territories, worlds, or singular determinations that are "more" or "less" intensely rhizomatic or arborescent, never simply one or the other: "A multiplicity has neither subject nor object, only determinations, magnitudes, and dimensions that cannot increase in magnitude without the multiplicity changing in nature" (*TP* 8).

Just as D&G deploy rhizomatics to limn the contours of this intensity, so Deleuze earlier in his career had used the language of the vegetable *psukhe*, lifted directly from Aristotle's *De anima* and its thinking about the various functions of the "soul" in animals, plants, and humans. Deleuze lays out the vegetable soul of rhizomatics in *Difference and Repetition*: "A soul must be attributed to the heart, to the muscles, nerves, and cells, but a contemplative soul whose entire function is to contract a habit [a deterritorializing line that must extract a territory, a 'world,' a form of content]. This is no mystical or barbarous hypothesis. On the contrary, habit here manifests its full generality: it concerns not only the sensor-motor habits that we have (psychologically), *but also, before these, the primary habits that we are; the thousands of passive syntheses of which we are organically composed*" (*DR* 74, my emphasis). This attribution of a "soul" and the power of "contemplation" to mechanical and chemical bodily processes may seem a little odd—and as I noted above, precisely the sort of thing that will get you called a "naive vitalist"—but Deleuze is here in the 1960s clearly working with a prerhizomatic understanding of the vegetable soul as that power of becoming that simultaneously striates a territory or a world. He defines such vegetable contemplation not as understanding or thinking but insists that "to contemplate is to draw something from": "contemplating—that is to say, in contracting that from which we come" (*DR* 74). He goes on to explain, "What we call wheat is a contraction of the earth and humidity, and this contraction is both a contemplation and the auto-satisfaction of that contemplation. By its existence alone, the lily of the field sings the glory of the heavens, the goddesses and gods—in other words, the elements that it contracts in contracting. What organism is not made of elements and cases of repetition, of contemplated and contracted water, nitrogen, carbon, chlorides and sulphates, thereby intertwining all the habits of which it is composed?" (*DR* 75). Inaugurating an argument that D&G will later expand in *A Thousand Plateaus* and *What Is Philosophy?*, here in *Difference and Repetition* Deleuze insists that humans, rocks, and lilies of the field are all composed of countless subroutines, passive syntheses that make "agency" or organic identity possible: "Underneath the self which acts are little selves which contemplate and which render possible both the action and the active subject. We speak of our 'self' only in virtue of these thousands of little witnesses which contemplate within us: it is always a third party that says me" (*DR* 75), he concludes, just as it is always a derived entity that is an organism. Again, there is *a* life before there is *my* life or *your* life.

This then is the deceptively simple upshot of Deleuze's *Difference and Repetition*: "Difference inhabits repetition" (*DR* 76) just as the swarmed field of individuation in its multiplicity traverses any "individual" (which is itself a form of habit, made of myriad other habitual structures); it's difference all the way across the "thousands of habits of which we are composed" (*DR* 78). And this leads us to *Difference and Repetition*'s rendition of "n−1," the rhizome's formula in *A Thousand Plateaus*: "below the level of active syntheses, the domain of passive syntheses which constitute us" (*DR* 79). Our "habits" or "contemplations" contract a territory for living out of the chaos of being: "Contemplating is creating,"[26] they insist in *What Is Philosophy?*, just as "habit is creative. The plant contemplates water, earth, nitrogen, carbon, chlorides and sulphates, and it contracts them in order to acquire its own concept. . . . We are all contemplations, and therefore habits. I is a habit. Wherever there are habits there are concepts" (*WP* 103). Creation, as I've intimated all along in this chapter, is a striating operation more so than a "freeing" operation—or, to put it more exactly, one must constitute a territory to diagnose where lines of re- and deterritorialization might exist. So, for example, philosophical concepts—the "territory" of philosophy—are neither inherently liberating nor freeing but need to be configured, deployed, and analyzed on their specific terrain or territory. As D&G insist in their final work, "philosophy is reterritorialized on the concept. The concept is not object but territory. It does not have an Object but a territory" (*WP* 101). Hence, philosophy creates concepts by marking out or creating territory for thinking; as philosophy's mode of becoming (striating territory as individuated emergence), creating concepts does not primarily undermine preexisting reified ideas, to free them to an "outside" of some kind. Creating concepts is striating new territory for thought.

Ontologically, one might sum up D&G's "n−1" method like this: attend not so much to organisms but to the territories created by the singularizations of the constant rhizomatic flows and subroutines that inhabit and constitute any given being; attend to the living thing not as an autonomous entity striving to remain intact in its world but as a singular swarm of passive syntheses inhabiting a territory made from lines of flight. Life is not housed exclusively in living animal or human beings and their potential-saturated worlds; rather, life names a distributed, inorganic swarm of emergent singularities that has often gone by the name of the "vegetable soul."

Politics: Capitalism and Rhizomatics

Stated as such, however, one might wonder whether the contemporary sociopolitical assemblage that most intensely renders this rhizomatic lesson of "n−1" is not vegetable life at all, nor even becoming-animal, but advanced global capitalism, where it's all global flows and arising singular niche markets all the time. Neoliberalism has proven to be quite adept at attending to such ahuman flows (of capital, desire, profit), and one might note in passing that contemporary capitalism infamously cares little about an old-fashioned humanist organicism (with workers now known simply as "human capital"). Today's cutting-edge capitalists want as little as possible to do with bricks-and-mortar entities, living or nonliving (other than as consumers or service providers). As anyone in the finance sector will confide after a martini or two, you don't primarily want to invest in preexisting *things* or *people*; you want to ride emergent *flows* into new products and markets, new territories: n−1.

In both theory and practice, then, the giddy philosophical thrust of D&G's work on the rhizome tends to deflate when the conversation turns specifically to human political power and to capitalism—which, as they remind us, is unique as a sociopolitical organizing principle insofar as capitalism encourages and benefits from deterritorialized flows (of money, labor, flexibility) rather than constantly trying to ward them off (as the older political sovereign and state forms consistently did). Many critics, including Slavoj Žižek, have argued that it's precisely their trumpeting this world of rhizomatic flow that makes D&G's work function as ideological cover for contemporary capitalism. Simply put, it can seem that D&G's insistence on rhizomatic flows rather than organisms provides a kind of philosophical warrant for naturalizing the practices of contemporary capitalism.[27] In a disciplinary world of rigid political segmentation (say, monarchy or fascism) such an ontological stance emphasizing grounding flows has an obvious progressive or resistant political valence. However, in a neoliberal world that dreams not of rigid obedience to accepted norms but of endless flows (or maybe finance capital's world dreams of rigid obedience to the idea of endless flow), the "n−1" tools of rhizomatics can seem less than revelatory: embedded within any political object or practice, we should learn to see swarms of financial investment. (Simply put: "Follow the money"?) Politically, the "n−1" methodology would seem to be less the one that could free lives (human or otherwise) from lines of social segmentation by capital and more to describe the ways we are interpellated by this new form of rhizomatic control.

Deleuze's "Postscript on the Societies of Control" constitutes a brief exploration devoted to the question of what comes after the rigid segmentations of Foucault's disciplinary society. Foucault's vision of the present and future, Deleuze insists, does not foresee more discipline, the increasing enclosure of subjects within stifling institutions. Rather, the present and future are characterized by the birth of new forms of power—more open, flexible methods aimed both at controlling individuals and gaining control over large aggregates like populations, sexuality, health, and even life itself. The late Foucault organizes these emergent regimes under the rubric of "biopower," but "control" is the related name that Deleuze appends to these lighter, more effective and more diffuse, methods of subject production in the present and future—ones that take as their target not this or that practice or institution but the very core practices of human "life" itself (say, sexuality, identity, or health). Control, in other words, names political rule by decentralized, rhizomatic means.

Importantly, Deleuze reminds us that Foucault's texts on the disciplinary society constitute historical work on the conditions that led up to the present, rather than representing an exhaustive analysis of contemporary social conditions:

> Foucault has brilliantly analyzed the ideal project of these [disciplinary] environments of enclosure, particularly visible within the factory: to concentrate; to distribute in space; to order in time; to compose a productive force within the dimension of space-time whose effect will be greater than the sum of its component forces. But what Foucault recognized as well was the transience of this model: it succeeded that of the societies of sovereignty, the goal and functions of which were something quite different (to tax rather than to organize production, to rule on death rather than to administer life); the transition took place over time, and Napoleon seemed to effect the large-scale conversion from one society to the other. But in their turn the disciplines underwent a crisis to the benefit of new forces that were gradually instituted and which accelerated after World War II: a disciplinary society was what we already no longer were, what we had ceased to be.[28]

Foucault's vision of the present and future, Deleuze argues, is not one that foresees the increasing institution of normative limits (which might then be transgressed by bold, resistant agency). Rather, the present and future are characterized by the birth of new forms of power—more open-ended, flexible methods aimed both at controlling individuals and gaining control over large, difficult-to-define aggregates. "These are the *societies of control*, which are in the process of replacing disciplinary societies. 'Control' is the name [William] Burroughs proposes as a term for the new monster, one that Foucault recognizes as our immediate future."[29] Although Deleuze here states that he takes the word *control* from Burroughs, he could just as easily have taken it from *Discipline and Punish*, where Foucault describes the "swarming [*l'essaimage*] of disciplinary mechanisms" that eventually reaches a tipping point in biopower, wherein "the massive, compact disciplines are broken down into flexible methods of control."[30] Or perhaps Deleuze was influenced by Foucault's sense of the word in a 1984 interview, when he states: "The control of sexuality takes a form wholly other than the disciplinary form that one finds, for example, in schools."[31]

When power produces control in Foucault's late work, that control is constituted less by a disciplinary mastery over specific individuals or populations than it is characterized by power's infiltrating ever-more-rhizomatic parts of the socius, finally saturating even the subject's relation to herself: the discourse of identity (or self-identity) becomes a means of control under the regime of biopolitics. Of course, discipline had its own political investments in subjective identity. But people have long been able to resist or reinscribe the brand of control deployed by disciplinary forms of identity: you can escape being a soldier, a wife, or a factory worker by fleeing from the army, the marriage, or the job. But it's much harder to escape the brand of control deployed within the rhizomatic biopolitical field. Take Foucault's primary example of "sexuality": whether you want one or not, everybody has got a sexuality, which is to say a complex of self-understood sexual investments that are not necessarily tied to (or bound by) our more segmented, disciplinary roles. Not everyone has a shared disciplinary identity (mother, student, cop), but everyone does have something like a sexual identity. And this remains the case even if one attempts to resist the practices and discourses of sex altogether: asexuality is still a sexuality.

In other words disciplinary power maintains control primarily through a training grid that is deployed and reinforced in myriad institutions, while postdisciplinary control largely functions through reorganization of the field in which any individual human's agency takes place. In Deleuzean parlance, control reorganizes the territory of the human assemblage through a political shift: from training the subject's actions at various institutional worlds (the subject in relation to the clinic, the family, the school, the army, the workplace) to working primarily on the subject's relation to herself, which is at stake virtually everywhere, at work all the time in thousands of tiny subroutines. Subjective identity is no longer a bulwark against political striation (identity politics vs. state mandates), but subjective identity constitutes the primary coemergent *territory* of politics in the society of control. Conservative Americans no longer merely *vote* Republican; they *are* Republicans. Being a Democrat or a Republican is no longer primarily projecting or subscribing to a vision of the world; it's a matter of inhabiting a biopolitical territory.[32]

Paradoxically, Foucault shows us that the more open and flexible any given regime of biopolitical practice (the more rhizomatic its processes, not clumsily contained in a centralizing organism or organization), the more effective the means of control can become. As Foucault suggests in his lecture series *The Birth of Biopolitics*, for example, global neoliberal capitalism can produce a society of control much more efficiently than a society of sovereignty or a society of disciplinary surveillance. Neoliberalism is a much more effective means of social control than sovereignty or discipline ever was precisely because of its supposed commitment to "openness" and flexibility. Biopolitical control, like the disciplinary regime out of which it arises, can hold better and saturate a greater area of the socius when its grip is not merely negative (repressive) but positive (enabling) as well. Political power holds sway not because it's a limiting nor enabling phantasmatic world but because it's an emergent territory.

Foucault sums up this form of nondisciplinary control in the last lecture of his *Birth of Biopolitics* series in 1979:

> You can see what appears on the horizon of this kind of analysis is not at all the ideal or project of an exhaustively disciplinary society in which the legal network hemming in individuals is taken over and extended internally by, let's say, normative mechanisms. Nor is it a society in which a mechanism of general normalization and the exclusion of those who cannot be normalized is needed. On the horizon of this analysis we see instead the image, idea, or theme-program of a society in which there is an optimization of systems of difference, in which the field is left open to fluctuating processes, in which minority individuals and practices are tolerated, in which action is brought to bear on the rules of the game rather than on the players, and finally in which there is an environmental type of intervention instead of the internal subjugation of individuals.[33]

With changes in the dominant modalities, practices, and targets of power, so too are the forms of social control transmogrified, made lighter, more intense, simultaneously more individual and more global. The individual's identity becomes the pivot of power for the late Foucault (just as the site of training in institutions had been the primary pivot and control mechanism for discipline). Our relation to ourselves (our "life" and our "world" of possibilities) constitutes that place where we are most intensely connected to biopower's modalities of social control; but precisely because of that fact, the territory of the self is also a privileged place where we might collectively respond to biopolitical control. While not an iron cage to be escaped, more intensely inhabiting the territory of our selves is just as surely not going to free us from the contemporary biopolitical apparatus, insofar as the subject and its world of ownmost possibilities constitutes the pivot of that contemporary apparatus.

And this clash is as true on the territories of the political as on the ontological ones, at least as far as Deleuze is concerned (Foucault remains largely allergic to ontology-talk). The slogan that Deleuze associates with Foucault's work on power—"Resistance comes first"—could just as easily be appended to his own work, with and without Guattari. As we have seen, it is an intense concern for a very robust sense of "life" that pushes Deleuze and Guattari's work well beyond considerations of human lives into an engagement with a global mesh of life-forms or "lifescapes." But for better or worse, resistance within this rhizomatic mesh doesn't guarantee anything politically, or maybe even ontologically. It's just a fact of life—or, rather, it's a fact of "a life," which by definition is a territory that resists capture (by death, for example). In a very late capitalist territory made up of lines of flight, strategies of deterritorialization remain as available to anarchists and union organizers as they are to army counterinsurgency experts and neoliberal finance executives.

In this context it seems important to recall the specific political note on which Deleuze ends the first section of his "Societies of Control" essay: "There is no need to fear or hope," he concludes, "but only to look for new weapons." In other words, for Deleuze the political upshot of power's rhizomatic status does not rest in some ideological unmasking or knowledge about the state of things in the world (our territory is liberating us, or our territory is constraining us; we should be either hopeful or fearful) but in that the primary usefulness of rhizomatic politics accrues to its *diagnostic* functions.

Rhizomatics diagnoses the territory: which is to say, the striating functions of our historical territory need to be named and explained, limned out, if we are to have any hope of looking for new concepts to combat them or to activate them otherwise. And Deleuze's primary homage to Foucault in this diagnostic project comes in affirming Foucault's historicism. Deleuze's may be a largely ontological project (every formation is built of lines of flight—an ether rather than an architecture), but his analysis of contemporary capitalism remains trained on a precise political diagnosis of the current configuration of this territory (how striation takes place within contemporary capitalism, what lines of flight might be available—where escape is possible, what lines are closed off). What freed people within the disciplinary human territory fifty years ago (maybe identity politics, maybe transgressive sexuality, maybe a commitment to indeterminacy and openness) will not be able to free them quite so unproblematically in the present.

The skeptic will counter that Walmart takes territory from mom-and-pop stores in precisely the rhizomatic way that switchgrass overtakes a meadow— which is of course correct; but as a diagnosis, it doesn't seem to follow that rhizomatics is an inherently dangerous political notion, a naturalizing beard for advanced capitalism (any more than the sense that transgressive sexuality or identity politics inherently injures or disrupts the workings of capitalism).

As a diagnostic discourse, rhizomatics is neither good nor bad. Rhizomatic *effects* most certainly can be judged according to that ethical yardstick (cancer spreads rhizomatically—and that's very dangerous indeed for human life); but the transversal field in which those effects happen is beyond good and evil. (To continue the example, you don't combat cancer because cancer is "bad" but because its effects are devastating for a particular life; and no matter how you feel or what you know about cancer, in order to fight it, you need to be able to diagnose it and to map how it spreads to construct and expand its territory: knowing that it's bad is of very limited use in combating its effects.)[34] This is perhaps the primary cultural lesson of the rhizomatic control society in Deleuze's sense: however dangerous the cage, it's still built from lines of flight. Just as the vegetable life of emergence flows through the organisms that seem to contain or define life, resistance is the force on which power feeds; resistance is what power seeks to canalize or use, just as the organism seeks to canalize the chaotic forces of life. Far from suggesting the nihilist night in which all cows are black (power is everywhere, even in the rhizome, so resistance is futile), Deleuze's primary lesson in the "Societies of Control" essay is that hope or despair is ultimately beside the point.

Deleuze's sense that new weapons have yet to be invented suggests that new territories for living have yet to be configured (as they say, "the people are missing"); but such an experimental construction of a territory is at the end of the day the entire political and ethical terrain of a life: not something to be projected, understood, mourned, celebrated, or lost but something to be done. A life is an art of effects, which is not an escape from politics (with Deleuze leading us "out of this world") but an intense, consistently experimental engagement with constructing the world as an emergent territory for living. "Follow the plants" doesn't suggest that everything becomes an indifferent plant any more than becoming-animal invites all things to be weaselly. Rather Deleuzoguattarean thought traverses an emergent engagement with the vegetable soul, which connects everything to everything else, an ecological mesh that forms an intense territory for living that goes all the way across the rhizome.

In the end *Plant Theory* will perhaps have argued nothing other than this: the vegetable *psukhe* of life is a concept or image of thought that far better characterizes our biopolitical present than does the human-animal image of life, which remains tethered to the organism, the individual being with its hidden life and its projected world. If we really are approaching the end of the world—with catastrophic climate change, ecological disaster, and maybe even human extinction looming very close by—it might be time to start diagnosing the world not as a static or dynamic backdrop for the myriad (im)possibilities of individual lives but as the ecological territory that cuts across all strata of life as we've known it, life as primarily defined rhizomatic territories, which is to say by the practices of emergence and transformation. And maybe in the end this only brings us back to the beginning. As Simondon writes of ancient Western thought in his *Two Lessons on Animal and Man*, "What appears to be quite clear . . . is that the human soul . . . is not considered as different in nature from the animal soul or the vegetal soul. Everything that lives is provided with a vital principle, the great dividing line passes between the reign of the living and the non-living much more so than between plants, animals, and man."[35] What we share with other entities, in other words, is less an abstract world of possibility than a territory for living. And as territories for living continue to disappear over the next century, maybe we'll begin to see rhizomatics less as a kind of metaphor for a better human world (freer or more open, more like the plants or the animals) and more as the operating system that traverses all facets of a life.

CODA

What Difference Does It Make?

AS I NOTED IN MY PREFACE, while the category of life certainly has been explored and expanded in the posthumanities-theory era, it seems to have been enlarged incrementally to take into consideration (some) animals—in point of fact, mostly those so-called charismatic animals who meet the "like us" standard set out in the foundational animal-studies work of Peter Singer and Tom Regan.[1] But that's about where the big-tent theory of life often ends, even though, as biologist Stephen Hubbell notes, "The proportion of the world's species that are charismatic organisms is really tiny. From a biomass point of view, this is a bacterial planet. It's a very parochial view to assume that we should care only about elephants and zebras."[2] If you really want to talk about life on this planet, by far most of the action is in the single-cell and vegetable kingdoms, which comprise numbers that dwarf all combined forms of animal life (including us) on a truly staggering scale: on land the vegetable kingdom makes up 99 percent of the total biomass, and the ocean floor is home to ten million trillion microbes for every human on the planet.[3]

But of course animal studies draws attention to charismatic animals for what I think are very good political reasons. The problems faced by such animals today are vast and pressing: farm animals are routinely caged in horrendous circumstances, then slaughtered in record numbers, while their sibling animals in the shrinking "wild" are undergoing a bona fide extinction event: by most estimates, each day we lose at least twenty-five animal species to the eternal night of extinction.[4] Given these tragic facts, it seems almost irresponsible for us to get caught up in a theoretical shell game of what counts and doesn't count as life that deserves political recognition or protection. As a concrete example of how "politics trumps" in animal studies, Cary Wolfe ventriloquizes Judith Butler's notion of "precarious life" like this: "'Not everything included under the rubric precarious life'—plants, for example—warrants protection from harm."[5] (In context, Butler here is actually talking about abortion rights— the fetus, she argues, is not entitled to "precarious life" political protections; it's Wolfe's interpolation that refers to plants.) The point, however, is that the pressing political questions of "protection from harm" seem to stifle, or at least background, the theoretical question of what counts or doesn't count as a viable form of life, precarious or otherwise.

This is ironic, insofar as on this account, nonanthropomorphic or posthumanist discourse seems to originate (rather than conclude) with the question of human rational choice. In his closing discussion of Derrida in *Before the Law*, from which I quoted a snippet in my Preface, Wolfe argues that "this act of selection and discrimination, in its contingency and finitude, is precisely what opens it [Derridean hospitality] to the other and to the future. This is why discrimination, selection, self-reference, and exclusion cannot be avoided, and it is also why the refusal to take seriously the differences between different forms of life—bonobos versus sunflowers, let's say—as subjects of immunitary protection is, as they used to say in the 1970s, a 'cop out.'?"[6] If I'm reading this right, Wolfe is suggesting that in Derrida there is no opening to the other nor any future (in context, no "hospitality") without the inaugurating act of human political choice ("selection and discrimination" among life-forms that deserve recognition and protection, and those that don't). This seems to me an astonishing claim. Even before the "cop out" 1970s, way back in the "groovy" 1960s, Derrida forces us (specifically, via 1966's "Structure, Sign, and Play") systematically to trouble this category of human choice. Indeed, ethically or politically, it would seem that the question of "life" would constitute what Derrida dubs a "region of historicity," wherein "the category of choice seems particularly trivial."[7] If we are, in fact, to rethink the question of life from the ground up, it seems to me that we can't go into the discussion already having chosen what counts as life and what doesn't—bonobos yes, sunflowers no.[8] It's likewise unclear to me that "immunitary protection"—which is to say, enforcing a nation-state-based human legal code in the name of animals (to the exclusion of plants or other forms of life)—is the sine qua non for responding to something like the precariousness of life today, across the spectrum.

I'm of course not calling for the arrest of people who mow their lawn, any more than Wolfe would suggest locking up meat eaters. In either case (in terms of plant or animal life) I'm merely pointing out how it's not clear that one can jump immediately to the legal or the political from these kinds of philosophical discussions about what life is or how life works. Humans will inevitably end up where they already are—in the territory of necessary but difficult political choices; but it does seem that the Derridean "trial of the undecidable" constitutes a crucial stage of such human political calculation. Given recent research on plant intelligence, for example, it's becoming clearer that most if not all of the things that are said when it comes to protecting animal lives from harm—animals have senses, can size up their territory, respond to threats, communicate, compete, remember—can likewise be said of plant life; but how (if at all) does that translate into a human "political" stance toward plants? Is the analogy to the human (how much any given entity is like or unlike us) still the only kind of "posthuman" criterion we can conjure for thinking about viable forms of life? Surely we can't simply decide to forgo eating plants as well as animals (as there's then nothing left for us but the deep ecologist slogan, "Do the Earth a favor and kill yourself"). But it likewise may turn out that in the future, eating "this, not that" will no longer offer humans a moral high ground.

If nothing else, taking vegetable life seriously imbricates us inexorably with other forms of life in such a way that our most intimate biopolitical actions (what we consume, for example, or what species we might want to help protect) can no longer seem unproblematic: I choose to eat only plants, or things without a central nervous system, and that makes me a better, more responsible person. Maybe, maybe not. On the plane of immunitary protection, we need to protect bonobos, of course, but the global consequences for all modalities of life may be greater if we don't do something to protect plants. Take, for example, the ocean's phytoplankton, which produce 40 percent of the world's oxygen (while processing similar volumes of carbon dioxide) and constitute the baseline for the food chain in the seas. These tiny plants are disappearing—their numbers are down 40 percent since 1950 because of rising sea-surface temperatures—and once they're gone, the oceans may quickly become dead zones of extinction.[9] And if 65 percent of the planet's surface becomes a toxic soup and there's 40 percent less oxygen in the air, then both the bonobos and the sunflowers won't be far behind. And if that's not a reason for even the most skeptical human to care about plants or to think about life and its protection in a more distributed way, then I don't know what is.

Plant Theory likewise should have other biopolitical consequences. At present, companies like Monsanto are legally allowed to patent—literally to own—some forms of plant life, but courts have balked (quite reasonably, it seems to me) at offering corporations such wide-ranging ownership over human life. Maybe an enhanced sense of the territory of life would cause a remapping of this divide, rethinking a private entity's ability to own plant life? Here I seem to be more in agreement than disagreement with Wolfe, in his attempt to bring together biopolitics and animal studies: he sees the common project "not just in resisting the formative dispositifs of modern biopolitics in their most brazen form but also in articulating with other dimensions of political resistance, such as opposition to the commodification and private ownership of life in the services of late capitalism. In this light, paying attention to nonhuman animal life has the potential to actually radicalize biopolitical thought beyond its usual parameters."[10] Agreed, but the mandate to "radicalize biopolitical thought beyond its usual boundaries" would, I'm arguing here, likewise extend to a consideration of vegetable life: an engagement that would indeed take biopolitical thought beyond its current (animal-friendly) incarnations.

In particular if we're concerned about the "private ownership of life in the services of late capitalism," as Wolfe is, we really don't want to turn our attention away from the vegetable kingdom, as much of the North American food chain—around 90 percent of all crops grown in the United States—is owned by a handful of powerful seed and pesticide manufacturers. We know the horrific story of animals being made to live on factory farms or left to die in dwindling habitats, but the plant kingdom has its own frightening story to tell about the biopolitical colonization of "life"—about the food chain itself being altered, with plants made to live differently through genetic modification. In thinking a more robust biopolitical sense of engagement, we might highlight the fact that "the US Patent and Trademark office grants more than seventy thousand patents a year, about 20 percent of which involve living organisms,"[11] and the numbers on genetically modified foods are particularly overwhelming: "Based on USDA survey data, HT [herbicide tolerant, genetically modified] soybeans went from 17 percent of US soybean acreage in 1997 to 68 percent in 2001 and 94 percent in 2014. Plantings of HT cotton expanded from about 10 percent of US acreage in 1997 to 56 percent in 2001 and 91 percent in 2014. The adoption of HT corn, which had been slower in previous years, has accelerated, reaching 89 percent of US corn acreage in 2014."[12]

Indeed, if you want to know what the future of patenting and financializing life at all levels looks like, forget animals and humans, and gaze upon the vegetable kingdom—where, for example, most US farmers are de facto required to buy their seeds from Monsanto each and every year, because to save seed from year to year would be to usurp Monsanto's intellectual property. As Peter Carstensen, from the University of Wisconsin Law School, puts it discussing "the image that Monsanto likes to use" when they talk about "ownership" of the seeds they force farmers to buy: Monsanto "compares a transgenic seed to a rental car: when you've finished using it, you return it to the owner. In other words, the company doesn't sell seeds, it just rents them, for one season, and it remains the permanent owner of the genetic information contained in the seed, which is divested of its status as a living organism and becomes a mere commodity. Finally, farmers became users of Monsanto's intellectual property. When you realize that seeds are the basis for feeding the world, I think there are reasons to be worried."[13] Likewise, when and if the patenting of human life becomes a reality, it will not emerge full-blown at the level of the organism (as that would be slavery, or at least indentured servitude) but at the bacterial or microbial level, where genetic manipulation and clinical applications with certain kinds of friendly bacteria are at the forefront of health research. In the future you and I may still own our bodies, but unless something changes relatively quickly, we'll do so as the Monsanto farmer owns his field: we'll be completely dependent on serial purchasing of expensive patented materials to keep the enterprise alive. (Of course, maybe that future is already here, given the high saturation of pharmaceuticals in many human and animal lives.) But in any case, the mesh of life and neoliberalism at the molecular level is morphing very quickly indeed, and we need a similarly robust biopolitics, one that moves at the level of life itself, to diagnose and respond to it.

Whatever specific debates we might want to revisit with an expanded sense of life in the tool kit, recent research on the Anthropocene has confirmed one thing we can all agree upon: Homo sapiens hasn't been doing the vast majority of other life-forms any favors over the past ten thousand or so years; so going forward, we should at least be clear that life is a mesh of emerging forms, not a competition among preexisting organisms. The question for politics in the biopower era is not really what we humans *should* do (as if our good intentions could immunize us against inflicting more harm), but as Foucault suggests, maybe we need to pay closer attention to what our doing does: as Foucault writes, "People know what they do; they frequently know why they do what

they do; but what they don't know is what what they do does."[14] If individual organisms are no longer the only forms of life that matter in terms of biopolitics,
this focuses our attention less on our own private, hidden, interior dilemmas
("Do I have the veggie burger or the fish sandwich?" "Do I support the sunflowers or the bonobos?") and more globally on "what doing does": which is to
say, examining how the mesh of life is altered by x or y practice, rather than se-
curing the best theoretical or epistemological ground for our political actions.
Life is an interlocking assemblage of forms and processes, a series of doings, as
Deleuze and Guattari insist; it is not a hidden world possessed by an individual
organism. This is the primary biopolitical provocation of *Plant Theory*.

Or in more academic parlance, you might think of the upshot of *Plant
Theory* this way: there's a common discursive move that, for better or worse,
has become the signature of "theory," especially the politics of theory, over the
past few decades. It has for a long time seemed common sense that, for example, you must first have a definition or concrete sense of "woman" before
you can have a feminism based on that secured identity. Or you must rely on a
bedrock definition of "the black subject" or "the queer subject" before you can
have a minoritarian movement based on race or sexual orientation. You'd need
to know who the proletariat is and what utopia you're working toward to have
any viable Marxism going forward. You'd need to know what film is and isn't
before you could produce something like visual-studies critique, or you'd need
rigorously to define what nature is before an ecological criticism can argue that
nature requires immunitary protection from the encroachments of a polluting,
destructive human culture. These things seem only common sense.

But humanities theory has, for better or worse, systematically undermined
this common sense over the past several decades. The bedrock discursive maneuver that Butler's feminist queer theory has in common with the antiessen-
tialist race criticism of Henry Louis Gates—or the poststructuralist Marxism of
Fredric Jameson, the visual culture work of D. N. Rodowick, or even the ecological work of Timothy Morton—is precisely this signature move: for the discourses of feminism, queer theory, racial and ethnic critique, Marxism, visual
culture, or even environmentalism to get anywhere in the future, they must
give up (rather than stubbornly hold on to) a strict definition of their subtending or defining "object."

That's suicide, you say: you need to have a clearly defined subject before
you can effectively enter any academic discourse, political discussion, or contentious social arena. But Timothy Morton sums up the other side of this ar-

gument, here discussing the project of "theory" that he undertakes in *Ecology Without Nature*:

> Some will accuse me of being a postmodernist, by which they will mean that I believe that the world is made of text, that there is nothing real. Nothing could be further from the truth. The idea of nature is all too real, and it has all too real an effect upon all too real beliefs, practices, and decisions in the all too real world. True, I claim that there is no such "thing" as nature, if by nature we mean some thing that is single, independent, and lasting. "Nature" is a focal point that compels us to assume certain attitudes. Ideology resides in the attitude we assume toward the fascinating object. By dissolving the object, we render the ideological fixation inoperative. At least, that is the plan.[15]

Morton argues a version of what all those others have argued before him, in the name of theory: one has to take the unproblematic object of "Nature" out of environmentalism in order for ecocriticism to be able to deal with the shifting challenges of our time. The idea that nature (or subjective identity, or the working class, or sexual orientation) is a preexisting thing, that's always been there and always remained the same until we recently threatened it in some way, doesn't help us in responding to the always-already-entwined nature of humans and their fellow forms of life, not to mention their immediate surroundings. Politically no less than epistemologically speaking, the problems surrounding any given topic change: the objects change, the targets change, so the strategies of response have to change. What we need is not so much a secure object to protect against all comers (that way lies both political and epistemological dogmatism) but a series of ways to adapt and respond to rapidly shifting unknowns, accelerating threats, new opportunities.

Contra Morton, though, I would argue that the primary practice or ethos of theoretical inquiry today is less one of de-essentializing or debunking (at this point, everybody knows that the dice are loaded) than of paying attention to the ways that questions are framed or posed—in Deleuzean parlance, paying attention to the way the concepts are coformed with certain conceptual territories. It's neither good nor bad but dangerous to forget that "man" or "woman" or "nature" or "animal" or any number of political-conceptual categories have a history, and there are substantial costs to forgetting that history in the present. The territory of politics and ethics can change fast: animal lives weren't much on the radar fifty years ago, and they certainly are today. Legal same-sex marriage was largely unthinkable just a few decades ago, at least in the United States. But such visibility, as Foucault reminds us in *Discipline and Punish*, can likewise function as a sort of trap. The questions remain open, and the final word on the discourses of politics and their subtending objects is that the commitments, entities, and practices driving the inquiry likewise have a history—they mutate. Hence the object of the discourse is not subtending (beneath the surface) at all, or at least not in the way we think it is (as the secure epistemological ground from which we can venture out politically or ethically). As Foucault shows us in *The Order of Things*, "life, labor, and language" (biology, political economy, linguistics) emerged as discourses that traversed the biopolitical territory of the human sciences in the nineteenth century and beyond; but, of course, these concepts and arenas of study won't always do that conceptual work. And as the human sciences' object ("man") begins to slide out of view, as the territory of biopolitics shifts, so do these discourses have to change and adapt.

As you'll no-doubt have surmised, this theory-style questioning or reframing of the conceptual territory is what I've tried to do here with the topic of life in the era of biopolitics—life caught between humans, animals, plants, the earth, objects, bacteria, whatever. I've tried here to subtract the sense that we know what life is (to think a bit more about the "bio" in "biopower"), not to further mystify life in the move that Foucault shows to be the founding of the human-centered biopower era but in order that we might reexamine how life functions (or might function otherwise) in our present. I come not to bury life (to steal it away into some unthematizable fog, as someone like Agamben does) but to bring it back to the surface of our debates and discourses around biopolitics. Life is the territory of biopolitics, and as such, we need to keep it front and center.

This is, however, not to say that any and all life—anthrax or the smallpox virus, for example—deserves equal political respect or protection from humans (a common critique of "vitalist" thinking, one leveled most strenuously these days against the work of Roberto Esposito).[16] The "gotcha" question usually goes something like this: If you have a robust, affirmative sense of life, if lots of stuff is said to be alive in an ethically compelling way, how do you decide to let some life-forms live and thrive, while making others (say, cancer cells or flesh-eating bacteria) die? If you give up the subtending and guiding object, human and/or animal life, how do you make discriminations or decisions about what counts as life that deserves protection? For example, Wolfe's concerns about taking vegetable life seriously tend to come down to what he sees as the heavy political costs of such an unfettered vitalism. As he wonders, "if all forms of life are given

equal value, then we face questions such as the following: 'Will we allow anthrax or cholera microbes to attain self-realization in wiping out sheep herds or
human kindergarteners?'?"[17] Quite straightforwardly, Wolfe asks, "Do we extend
[Derridean] 'unconditional hospitality' to anthrax and ebola virus, to SARS?"[18]
While these are obviously rhetorical questions (I'm not sure why you need a
warrant from Derrida to seek treatment for Ebola), I might begin responding to
them by pointing out that Derridean "unconditional hospitality," like the trace
structure, is not something we "decide" to extend or not: such ontological openness to the future (and to death) in fact structures the field of Derridean life,
as we saw in Chapter 3. I can contract Ebola or SARS because of the structural
openness of my individual body and the larger social field in which I live. And
simply deciding not to get cancer is not going to save me from it. Derridean
hospitality is not akin to a dinner party guest list ("you're in, you're out"), any
more than "writing" in Derrida is confined to what I'm doing when I compose
these sentences. Such hospitality, like the Derridean notion of life itself, is not
originated by nor reducible to human embrace or neglect of this or that form of
life. In terms of eating as much as living, the challenge is to account as fully as
possible for various forms of violence, not to renounce the violence of choice or
life altogether (as if that were even possible). Thinking robustly about life isn't to
say that all life is the same, nor is it a ploy to make you feel bad about your cancer treatments. It is rather to suggest the opposite—that an untainted moral high
ground is impossible when it comes to thinking about meshes of life. Undecidability complicates decision; it doesn't make decision impossible.

Humanities "theory" has over the past few decades developed a series of
sophisticated ways for dealing with these sorts of "what about anthrax?" objections, which I think are really questions about how you make discriminations in a flat field of immanence or multiplicity: how do you decide anything
in a field bereft of a subtending ground, a "God" or a "real" or an "outside"?
(Remember the old days, when if you said that history is always beholden
to discursive or cultural mediation, then you were denying the reality of the
Holocaust? Or if you said power was everywhere, you were suggesting that resistance was impossible? Remember when "There is nothing outside the text"
supposedly meant we all live in a book, or "becoming-animal" was acting like
a chimp?) Like resistance to power and recalling the specificity of historical
traumas, the political question of immunitary protection for (or from) x or y
life-form is indeed a pressing one for humans. Our biopolitical discussions will
inevitably end up on that political territory of judgment so deftly analyzed by

Timothy Campbell in his *Improper Life*, and I'm with Wolfe and animal studies on that point. But I'm simply trying to point out one of the substantial transaction costs of reducing the transversal immanence of "life" to the instrumental rationality of human political decision (to the political conundrum of worthy versus unworthy life). In other words, I'm wondering here about the hidden cost of silently passing along the sovereign, individual, "animal" territory for thinking about life and the organism that's been characteristic of the politics and economics of the biopower era.

In fact, such a calling out of what Foucault calls the "blackmail of the Enlightenment" (either submit to the reign of political calculation over life, or suffer the pure chaos of unfettered mass death) underpins Esposito's proleptic response to Wolfe's "What about anthrax?" concerns. In *Bios* Esposito quotes Deleuze's *Logic of Sense*: "We cannot accept the alternative: . . . either singularities already comprised in individuals and persons, or the undifferentiated abyss."[19] In point of fact this "all or nothing" position on life is not the one that Esposito holds in *Bios*, where he remains very close to the Deleuzean reading of Simondon that I articulated in Chapter 4: if life is a constant modality of emergent individuation (a harmonious territorial mesh of refrains or subroutines), then any "individual" entity or living thing is already a "trans-individual." This ontological point contains no particular policy implications—that you should "value" x or y life-form—but it does give us a picture of how something like life works (through constant emergence rather than a dialectical struggle of life versus death, us versus them). Esposito makes this point time and again, stressing that his is an attempt to think biopower "not over life nor beginning from life, but in life, which is to say in the biological constitution of the living organism";[20] to think biopolitics through "the development of individual or collective life. Or better: the moving line that runs from the first to the second, constantly translating the one into the other. As we know, it's what Simondon defines with the term and the concept 'transindividual.'?"[21] In short, Esposito's ontological claim is this: "As the human body lives in an infinite series of relations with the bodies of others, so the internal regulation will be subject to continuous variations" (188) in both the individual and collective (organic and political) register.

I opened *Plant Theory* with Foucault's reminder that it was plant life, rather than animal life, that was left behind at the ascendency of biopower. It should follow that the distributed territory of rhizomatic plant life will, hopefully better than the animal-organism territory for understanding life, offer us some tools to think our way forward in these, to use Morton's apt phrase, dark ecological times. As Karen Houle puts it succinctly, it may be that "thinking plant-thoughts shoves us, in a better way than thinking animal-thoughts does, toward the truth that the correct unit of analysis is not the individual, nor the dyad, but the assemblage. . . . Plant-becoming also radically re-imagines Life as that which can be accomplished not within a successfully-managed organic encasement of what a thing is (its being, its teloi, its progeny) but, as that which can happen by virtue of a certain unfaithful power of connectivity."[22] If plant life is largely sidestepped by biopolitics, it would seem that's the case primarily because when one imbricates vegetable life within the extant schemes of bio-power (which seems to have decided that life is a property of an individual or-ganism), one gets an entirely different version of life's territory. As Deleuze and Guattari put it in *A Thousand Plateaus*, "A multiplicity without an ancestor? It is quite simple; everybody knows it, but it is discussed only in secret. . . . Un-natural participations or nuptials are the true Nature spanning the kingdoms of nature. Propagation by epidemic, by contagion, has nothing to do with filia-tion by heredity. . . . The difference is that contagion, epidemic, involves terms that are entirely heterogeneous: for example, a human being, an animal, and a bacterium, a virus, a molecule, a microorganism. Or in the case of the truffle, a tree, a fly, and a pig. These combinations are neither genetic nor structural; they are interkingdoms" (*TP* 241–42). Life is not something that's owned by an organism, something hidden deep within it, to be protected against the outside at all costs; rather, life is the territory for the emergence of "interkingdoms," assemblages of heterogeneous processes. The animal territory for thematizing life, however important and apt it may be, tends to focus our attention on the biopolitical competition among individual organisms to the detriment of this robust sense of distributed, interconnected life. If, as Wolfe urges, "race and species must . . . give way to their own deconstruction in favor of a more highly differentiated thinking of life in relation to biopower,"[23] then so too must the Linnean kingdom divide that still bifurcates life (animal/vegetable) be subject to a deconstruction.

The End

And on the other side of the coin of life is, of course, a rethinking of death. To circle back to the Agamben between Derrida and Foucault with which I began in Chapters 1 and 3, recall that for Derrida death remains the ultimate in tragic pathos (the loss of everything, of the whole world—which is again to say death remains understood wholly within the orbit of the individual organism's tarrying with the negative birthed in Hegel and so decisively intensified in the nineteenth century). Whereas for Foucauldian biopower death is less a tragic drama of unredeemable loss than it is a question of medico-political administration—in his famous formulation of biopower it's a matter of managing populations rather than offering sovereign political decisions concerning individuals. And this *différend* concerning the primary sense of life and death is no longer merely an academic spat, as a whole baby-boom generation stares down this Foucauldian question of being made to live or left to die, a regime of life and death lying largely outside the drama of any individual's desires.

In the meantime, as we are beginning to see real signs of climate change leading to global mass extinctions as a terrifying backdrop to these more prosaic questions of long-term disability and nursing home care, it would seem there are substantial costs to continuing to see death in this individual, tragic, "animal" register—pivoting primarily on particular organisms and their life or death. One might in addition wonder concerning Derrida, is there an "each time unique, the end of the species"? Or does thinking life and death primarily through the individual organism largely preclude or at least deemphasize that kind of aggregate, species-level or "posthuman" thinking?[24] In the adolescence of the twenty-first century, staring down for the first time what looks like the inevitability (rather than the abstract possibility) of human extinction, maybe we are or soon will be experiencing a shift in our understanding and practices of life, a morphing every bit as radical as the one that Foucault suggests birthed life as we know it in the nineteenth century.

Perhaps it seems overly breathless to say that life itself is or soon will be undergoing a mutation, but it's probably worth remembering, in the genealogical spirit of Foucault, that the dominant human-sciences understanding of life has already changed radically at least twice in the past one hundred years (from the vitalism of the late nineteenth century, to the sense of life as information with the discovery of DNA, to the microbiological understanding of life and/as genetic manipulation in the present).[25] Likewise, I think we could agree that global climate disaster impacts conceptions of (human) life in a very different way than did the cold-war possibility of nuclear war leading to mass extinction. The apocalypse of so-called "mutually assured destruction," as Derrida himself pointed out several times, tended to preserve the tragic, hidden pathos of unthematizable death visited upon us at a moment's notice.[26] But global climate change will likely prove to be a different kind of life-and-death situation, with less sudden tragedy and more slow erosion of habitat and massive displacement of populations (about a quarter of the globe's human population presently lives in areas projected to be under water by the end of this century). The question of human life and death in the twenty-first century will likely prove to be an even more intensely and thoroughly administrative quandary.

This is likewise to affirm my friend Cary Wolfe's insistence that politics—human decisions concerning which lives count and which ones don't—will inevitably return in the end, as they must. Looking forward from the neoliberal consensus that seems to have formed in these early decades of this century, one has to wonder how humankind can respond to this large-scale administrative challenge in the era of the privatizing, small-government triumph that Deleuze and Guattari, Derrida, and Foucault all diagnosed toward the end of their lives—a situation dubbed "capitalist realism" by Mark Fisher (in the sense that it seems there's nothing but capitalism going forward: it has become the real).[27] Contemporary neoliberal nation-states are designed to protect and serve the very privatizing biopolitical pathos of nineteenth-century subjectivity, focused on the life of the individual organism. Capitalism's dialectical drama is fueled by individual desires (by consumption), and any "we" remains based on a prior thought of the desiring "I," in Hegel's famous formulation. So any kind of large-scale intergovernmental thinking about the future (which is specifically to say, high global taxes on assets and fossil fuels coupled with massive, coordinated global nation-states' spending on huge protection and renewal projects) remains economic heresy in the present; and it likewise augurs electoral suicide in the future for any politicians either bold or stupid enough to propose such governmental (rather than privatizing) solutions going forward. To paraphrase an observation of Fredric Jameson's, right now we're far better equipped to imagine extinction—the end of the world—than to imagine an alternative to global neoliberal capitalism.[28]

In the very end I would argue that all this commits us going forward to a diagnostic rather than a primarily moral (proper versus improper) critical discourse about life—to diagnosing what our "doing does" within an expanded mesh of life. In this vein Foucault provocatively writes, "I think the alternative to death isn't life but truth. What we have to rediscover through the whiteness and inertia of death isn't the lost shudder of life, it's the meticulous deployment of truth. In that sense, I would call myself a diagnostician."[29] I take this to mean that the airy biopolitical abstractions of hidden but protean life, facing off against an angst-ridden discourse of death, have largely played themselves out in our post-anthropomorphic world, where the questions of life and death (for all life-forms) are increasingly seen to be demographic, environmental, and global—related to the (admittedly grim) "truth" of the present historical situation rather than to that tragic "lost shudder" in which, as Thacker puts it, "life . . . really means life-for-me, or life-for-us."[30] Given this sobering portrait of the present and future, we may want to follow Foucault in thinking that death, like life, really *isn't* what it used to be; and that continuing to think about these biopolitical questions on the territory of the nineteenth-century (individual organisms and their phantasmatic worlds) may be less a philosophical or academic disagreement than it is quite literally a matter of life and death.

NOTES

Preface

1. If you run "Animals" through the MLA Convention program database (covering the years 2004–14), you get 186 sessions where the topic was discussed (either in a single paper or by an entire panel), but they're of course unevenly distributed in time—only seven sessions in 2007, for example, whereas there were twenty-eight presentations or panels that took up animality in 2012, thirty-one such sessions in 2013, and thirty-three in 2014.

2. Cary Wolfe, *Before the Law: Humans and Animals in a Biopolitical Frame* (Chicago: University of Chicago Press, 2012), 102.

3. The debate can be found on the Columbia University Press blog, http://www .cupblog.org/?p=6608.

4. These can be found at www.nytimes.com/2009/12/22/science/22angi.html; and www.nytimes.com/2011/03/15/science/15food.html.

5. *Summa Theologica, Third Part, Question 44, Article 4, Objection 1*, www.newad vent.org/summa/4044.htm.

6. So, bibliographic information for this long list: Michael Marder, *Plant-Thinking: A Philosophy of Vegetal Life* (New York: Columbia University Press, 2013); Richard Doyle, *Darwin's Pharmacy: Sex, Plants, and the Noosphere* (Seattle: University of Washington Press, 2013); Matthew Hall, *Plants as Persons* (Albany, NY: SUNY Press, 2010); Elaine Miller, *The Vegetative Soul* (Albany, NY: SUNY Press, 2002); Eduardo Kohn, *How Forests Think: Toward an Anthropology Beyond the Human* (Berkeley: University of California Press, 2013); Timothy Morton, *Ecology Without Nature: Rethinking Environmental Aesthetics* (Cambridge, MA: Harvard University Press, 2009); Claire Colebrook, *Death of the Post-Human: Essays on Extinction* (Ann Arbor, MI: Open Humanities Press, 2013); Jeffrey Jerome Cohen, ed., *Animal, Vegetable, Mineral: Ethics and Objects* (Washington, DC: Punctum, 2012); Theresa Kelley, *Clandestine Marriage: Botany and Romantic Culture* (Baltimore: Johns Hopkins University Press, 2013); Robert Mitchell, *Experimental Life: Vitalism in Romantic Science and Literature* (Baltimore: Johns Hopkins University Press, 2013).

7. Anthony Trewavas, "What Is Plant Behaviour?" *Plant, Cell and Environment* 32 (2009): 606–16.

8. Michael Pollan, *The Botany of Desire: A Plant's-Eye View of the World* (New York: Random House, 2002); Francis Halle, *In Praise of Plants*, trans. David Lee (Portland,

OR: Timber Press, 2011); See Sack's "The Mental Life of Plants and Worms, Among Other Things," www.nybooks.com/articles/archives/2014/apr/24/mental-life-plants -and-worms-among-others/; and Pollan's "The Intelligent Plant," www.newyorker.com/ reporting/2013/12/23/131223.

9. Daniel Chamovitz, *What a Plant Knows* (New York: Scientific American, 2013), 6, 137–38, 141.

10. See http://datagarden.org/7940/midisprout/.

11. Cited in Lisa Swanstrom's "Nature's Agents," www.electronicbookreview.com/ thread/electropoetics/natural.

12. There are of course a number of very important recent theoretical texts on "life" in and around Foucault, Derrida, and Deleuze and Guattari that I'll be building on here. The most incisive general discussions of the topic, to my eyes, are Eugene Thacker's *After Life* (Chicago: University of Chicago Press, 2010); and Roberto Esposito's *Bios: Biopolitics and Philosophy*, trans. Timothy C. Campbell (Minneapolis: University of Minnesota Press, 2008). Regarding Deleuze and Guattari, see Keith Ansell-Pearson's *Germinal Life: The Difference and Repetition of Deleuze* (London: Routledge, 1999); Claire Colebrook's *Deleuze and the Meaning of Life* (London: Continuum, 2011); and John Protevi's *Life, War, Earth: Deleuze and the Sciences* (Minneapolis: University of Minnesota Press, 2013). In terms of Derrida and the discourse of life see Richard Doyle's *On Beyond Living* (Stanford, CA: Stanford University Press, 1996); and Martin Hägglund's *Radical Atheism: Derrida and the Time of Life* (Stanford, CA: Stanford University Press, 2008). On Foucault see Nikolas Rose's *On The Politics of Life Itself* (Princeton, NJ: Princeton University Press, 2006); and Foucault's lectures on *The Birth of Biopolitics*, trans. Graham Burchell (New York: Picador, 2010). On Heidegger see David Krell's authoritative *Daimon Life: Heidegger and Life-Philosophy* (Bloomington: Indiana University Press, 1996); on Agamben and Heidegger see especially Timothy Campbell's *Improper Life: Technology and Biopolitics from Heidegger to Agamben* (Minneapolis: University of Minnesota Press, 2011).

13. Thacker, *After Life*, 6.

14. See Judith Butler, *Precarious Life: The Powers of Mourning and Violence* (London: Verso, 2006).

Chapter 1. The First Birth of Biopower

1. See James Gorman, "Animal Studies Cross Campus to Lecture Hall," *New York Times*, 2 Jan. 2012, www.nytimes.com/2012/01/03/science/animal-studies-move-from-the-lab-to-the-lecture-hall.html.

2. Michel Foucault, *"Society Must Be Defended": Lectures at the Collège de France, 1975–1976*, trans. David Macey, ed. Mauro Bertani and Alessandro Fortana (New York: Picador, 2003), 242.

3. As Foucault concisely puts it, the disciplinary body becomes "more obedient as it becomes more useful." Michel Foucault, *Discipline and Punish: The Birth of the Prison*, trans. Alan Sheridan (New York: Vintage, 1979), 138.

4. See, just as the most obvious examples of a vast and growing critical literature

concerning sustainability and the global meat industry, Robert Kenner's documentary film *Food Inc.*; and Jonathan Safran Foer's *Eating Animal* (Boston: Back Bay Books, 2010).

5. See Wolfe's extensive engagement with Foucault throughout *Before the Law*.

6. See, for example, Simon During's headnote to Adorno and Horkheimer's "The Culture Industry: Enlightenment as Mass Deception," in *The Cultural Studies Reader*, ed. Simon During, 3rd ed. (London: Routledge, 2007): "Adorno and Horkheimer neglect what was to become central to cultural studies: the ways in which the cultural industry, while in the service of organized capital, also provides for all kinds of individual and collective creativity and decoding" (32).

7. Donna Haraway, *When Species Meet* (Minneapolis: University of Minnesota Press, 2007), 59–60.

8. Nicole Shukin, *Animal Capital: Rendering Life in Biopolitical Times* (Minneapolis: University of Minnesota Press, 2009), 11.

9. Michel Foucault, *The Order of Things: An Archaeology of the Human Sciences*, trans. Alan Sheridan (1966; New York: Pantheon, 1970), 387. Hereafter cited in the text as *OT*.

10. On Foucault's dislike for polemic argumentation see his interview, "Polemics, Politics, and Problematizations," in *Foucault Live* (New York: Semiotexte, 1996).

11. In *Foucault Studies* 9 (Sept. 2010): 89–110, 90.

12. For an overview of Foucauldian approaches to animality see Chloe Taylor, "Foucault and Critical Animal Studies: Genealogies of Agricultural Power." *Philosophy Compass* 8, no. 6 (2013): 539–51.

13. Michel Foucault, *History of Madness*, trans. Jonathan Murphy (New York: Routledge, 2006), 19. Hereafter cited in the text as *HM*.

14. Re: 1870 and homosexuality see Foucault's 19 March 1975 lecture in Michel Foucault, *Abnormal: Lectures at the Collège de France, 1974–1975*, trans. Graham Burchell (New York: Picador, 2004); and Michel Foucault, *History of Sexuality, Volume 1: An Introduction*, trans. Robert Hurley (1976; New York: Vintage, 1980), 43.

15. Frank O'Hara, "Animals," www.frankohara.org/writing.html#animals.

16. Here I would also note that Foucault holds on to this early interest in animality and thought right up to the end of his philosophical itinerary. In his final lecture course, 1983–84's *The Courage of Truth*, trans. Graham Burchell (New York: Picador, 2012), Foucault devotes considerable time to the Cynics and their foreshadowing of the nineteenth-century romanticist discourse of desire as animality. As he argues in his 14 March 1984 lecture (a scant three months before his death in June), "There are still a great many things that could be said about this naturalness in the Cynics. This principle of a straight life which must be indexed to nature, and solely to nature, ends up giving a positive value to animality. And, here again, this is something odd and scandalous in ancient thought" (264). Foucault goes on to suggest the Cynics were decisive precursors for our modern understanding of subjectivity as transgressive, existential, "animal" authenticity: "This animality, which is the material mode of existence, which is also its moral model, constitutes a sort of permanent challenge in the Cynic life. Animality is a way of being with regard to oneself, a way of being which must take the form of a con-

stant test. Animality is an exercise. It is a task for oneself and at the same time a scandal for others" (265).

17. See http://en.wikipedia.org/wiki/Biomass_%28ecology%29.

18. See Trewavas's "What Is Plant Behaviour?" for an excellent overview of research in the area.

19. Aristotle, *On Plants*, in *The Complete Works of Aristotle: The Revised Oxford Translation*, ed. Jonathan Barnes, vol. 2 (Princeton, NJ: Princeton University Press, 1984), 820a10.

20. Of course the question of "life" has been enigmatic from the Western get-go, but I'd here highlight the difference that Foucault marks in the discourse of life "when philosophy becomes anthropology" in the early modern period in Europe. At that turning point, thinking takes human life (or, better, the enigmatic experience of life for humans) as its grounding movement. In Aristotle life is maybe hidden and disbursed—you can't think life "itself" or only through the examination of living things—but in whatever case, life in Aristotle is certainly not reducible to (a subject's) desire or experience. This idea is invented in the modern era. Likewise in medieval thinking, life is certainly enigmatic, but as Thacker points out in *After Life*, "inasmuch as the question of life was, for the Scholastics, less a question about what we would call biological or animal life, it increasingly becomes a question of some principle that is able to mediate between the purely material and the purely spiritual. In Scholasticism, Life is more than Being but less than God. . . . We should also remind ourselves of the inherent non-anthropomorphic tendencies of Scholastic heretical thinking in this regard. The question of the creature is not a metonym for any humanist crisis or the question of the human being's place in the world. The finitude of the creature is not defined by any anguish or privileging of the human" (155–56).

21. See the proceedings here: www.princeton.edu/ihum/events/plants/.

22. Marder, *Plant-Thinking*, 28, 30, 46.

23. For a fine book on vegetable life that doesn't fall prey to this logic, see Doyle's *Darwin's Pharmacy*. For more of my thoughts on Marder's *Plant-Thinking* see Jeffrey T. Nealon, review of *Plant-Thinking: A Philosophy of Vegetal Life*, http://ndpr.nd.edu/news/39002-plant-thinking-a-philosophy-of-vegetal-life/.

24. *The Archaeology of Knowledge*, trans. A. M. Sheridan Smith (1969; New York: Pantheon, 1972), 107, 120. Hereafter cited in the text as *AK*.

25. Recall the context in which Foucault infamously tags deconstruction as a "historically determined little pedagogy": such a generous ventriloquizing reanimation of "undecidable" objects and texts "inversely gives to the master's voice that unlimited sovereignty that allows it indefinitely to re-say the text" (*HM* 573).

26. In his *Implications of Immanence: Toward a New Concept of Life* (New York: Fordham University Press, 2006), Leonard Lawlor performs a similar analysis in reviewing Foucault on "the rupture that opened the modern epoch," a formation that Lawlor calls "life-ism": "When Bichat developed his definition of life, he placed life at a deeper level; he replaced nature with life as the ontological foundation. Life itself becomes the background or ground for all other oppositions. This shift to the level of ground implies that the traditional problems associated with the concept of life are pushed to the side, problems such as

the unity of life (vegetative versus inorganic), the opposition between finalism and mechanism, the conceptions of evolution. Instead, replacing being as well as nature, life becomes *ultra-transcendental*" (144). Lawlor and I agree on this Foucauldian diagnosis of the problem, but we disagree concerning Foucault's reaction to this "ultra-transcendental" notion of life that was birthed in the nineteenth century. For Lawlor, Foucault doesn't criticize or try to turn away from this murky transcendentalism of life but instead doubles down on it: "This fundamental blindness [of "life"], this inability to see, places powerlessness right in the very midst of the viewpoint of power understood as preservation-enhancement. . . . Only where there is blindness (and blindness always makes one uncertain), only where one (a life) is no longer visible to the general Panopticon, is resistance possible. With blindness, there can be no constant presence; there is always invisibility and thus freedom. In other words, from the points of impossibility (not the plenitude of the possible) comes the possible. Only by following the line of this powerlessness will we be able to twist free of the most current and dangerous form of metaphysics" (142). Thereby, it seems to me, Foucault's project becomes Derrida's, concerned less with concrete practices of freedom than with "twisting free" from the grip of the metaphysics of presence.

27. Giorgio Agamben, *Homo Sacer: Sovereign Power and Bare Life*, trans. Daniel Heller-Roazen (Stanford, CA: Stanford University Press, 1998), 9. Hereafter cited in the text as *HS*.

28. From Hanna Leitgeb and Cornelia Vismann, "Das unheilige Leben," *Cicero*, 4 Oct. 2010, www.cicero.de/salon/das-unheilige-leben/47168 (my translation).

29. Jacques Derrida, *The Beast and the Sovereign*, trans. Geoffrey Bennington, 2 vols. (Chicago: University of Chicago Press, 2011), 1:330. Hereafter cited in the text as *BS*.

30. For more on this see my *Foucault Beyond Foucault: Power and Its Intensifications Since 1984* (Stanford, CA: Stanford University Press, 2008).

31. From Foucault's *Remarks on Marx*, trans. R James Goldstein (New York: Semiotexte, 1991), 174.

32. From the interview "Truth and Power," in Michel Foucault, *Power/Knowledge: Selected Interviews and Other Writings*, ed. Colin Gordon (New York: Vintage, 1980), 121.

33. In fact, as Foucault makes abundantly clear in *The Birth of Biopolitics*, German ordo-liberalism of the 1950s placed its confidence in market capitalism precisely because it was seen as a potent corrective to Nazi statism.

34. Gilles Deleuze, "Postscript on Control Societies," in *Negotiations, 1972–1990*, trans. Martin Joughin (New York: Columbia University Press, 1995), 181.

35. Originally published in *La repubblica*, 8 Jan. 2004, 42. English translation at www.egs.edu/faculty/giorgio-agamben/articles/no-to-bio-political-tattooing/.

36. This linkage of animalization to the Holocaust is clearly important to Agamben (in fact sets up his entire project), but it is sadly nowhere to be found in Foucault's published works. It exists only as quoted in Hubert Dreyfus and Paul Rabinow's *Michel Foucault: Beyond Structuralism and Hermeneutics* (Chicago: University of Chicago Press, 1982) and seems to have been taken from a transcript of a discussion at Stanford after Foucault gave "Omnes et Singulatim" (in English) as the Tanner Lecture in 1979. Presumably, Foucault was asked by an audience member a question about the Holocaust,

and this was his improvised response (again, in English); or if this was indeed part of the original lecture, Foucault then excised this reference to the Holocaust from all subsequent published versions.

For his exact wording, Agamben is clearly translating—into Italian—the French rendering of Dreyfus and Rabinow's book *Michel Foucault: Un parcours philosophique*, trans. Fabienne Durand Bogaert (Paris: Gallimard, 1984); the original English in Dreyfus and Rabinow is clumsy and, as I say, not included by Foucault in any publication of the lecture. Specifically in context, Foucault is commenting on how "the importance of life for these problems of political power increases": "a kind of animalization of man through the most sophisticated political techniques results. Both the development of the possibilities of the human and social sciences, and the simultaneous possibility of protecting life and of the holocaust make their historical appearance" (Dreyfus and Rabinow, *Michel Foucault*, 138).

Foucault's somewhat inelegant English (all the verbs at the end of the sentences) is rendered much more lyrically by the French translation (which likewise moves the action—"what results" and "there appears"—to the beginning of the sentences): "il en résulte une sorte animalisation de l'homme au travers des techniques politiques les plus sophistiquées. Apparaissent alors dans l'histoire le déploiement des possibilités des sciences humaines et sociales ainsi que la possibilité simultanée de protéger la vie et d'autoriser l'holocauste" (202).

Compare Agamben's Italian rendering (clearly following the specific wording of the French translation, not the English original): "Ne risulta una sorta di animalizzazione dell'uomo attuata attraverso le più sofisticate tecniche politiche. Appaiono allora nella storia sia il diffondersi delle possibilità delle scienze uname e sociali, sia la simultanea possibilità di proteggere la vita e di autorizzarne l'olocausto" (*Homo Sacer: Il potere sovrano e la nuda vita* [Einaudi: Torino, 1995], 5).

37. See the chapter "Once More, with Intensity," in Nealon, *Foucault Beyond Foucault*.

38. Cited in Deleuze's "Desire and Pleasure," trans. Melissa McMahon, letter "G," www.artdes.monash.edu.au/globe/delfou.html.

39. Tim Dean, "The Biopolitics of Pleasure," *South Atlantic Quarterly* 111, no. 3 (2012): 477–96.

40. Giorgio Agamben, *The Open: Man and Animal*, trans. Kevin Attell (Palo Alto, CA: Stanford University Press, 2003), 15 (my emphasis).

41. Ibid., 16.

42. From Negri's conversation with Cesare Casarino, "It's a Powerful Life," in *Cultural Critique* 57 (Spring 2004): 151–83, 172.

Chapter 2. Thinking Plants with Aristotle and Heidegger

1. Jacques Derrida, *The Animal That Therefore I Am*, trans. David Wills (New York: Fordham University Press, 2008), 29.

2. Plato, *Timaeus*, 77a–c, trans. Benjamin Jowett, http://classics.mit.edu/Plato/timaeus.html.

3. See Trewavas's "Plant Behaviour" for the strongest scientific case for plant intelligence and behavior. In *What a Plant Knows* Chamovitz argues that plants have senses rather than intelligence and don't in either case feel pain. Tomato plants, for example, respond to injury not by immediate recoil of the damaged leaf but by signaling other leaves: "The leaf did not feel pain. The tomato responded to the hot metal not by moving away from it but by warning other leaves of a potentially dangerous environment" (68).

4. See especially Hall's *Plants as Persons* and Marder's *Plant-Thinking* for extensive documentation of the systematic philosophical abjection of vegetable life.

5. Marder, *Plant-Thinking*, 2.

6. There's obviously a large historical gap here, from Aristotle to Heidegger, but luckily Thacker's *After Life* contains excellent chapters on medieval and early modern conceptions of life in the West, traversing especially nimbly through Aquinas and Kant.

7. Much discussion of plants in Plato is nested within larger discussions of cultivation, a consistent metaphorics that paints the Socratic teacher as a kind of greenhouse keeper, tending to the young sprouts, while doing substantial weeding—"like a farmer who cherishes and trains the cultivated plants but checks the growth of the wild—and he will make an ally of the lion's nature" (*Republic* 9.589a–b). Much of that discourse boils down to this description of the method, which Socrates offers in *Phaedrus* 276e: "one employs the dialectic method and plants and sows in a fitting soul intelligent words." As Socrates expands elsewhere (responding in fact to Melitus's charges), "the right way is to take care of the young men first, to make them as good as possible, just as a good husbandman will naturally take care of the young plants first and afterwards of the rest" (*Euthyph.* 2d). Or see Plato, *Laws* 6.765e: "For in the case of every creature—plant or animal, tame and wild alike—it is the first shoot, if it sprouts out well, that is most effective in bringing to its proper development the essential excellence of the creature in question." For all these Plato texts and citations see MIT's Internet Classics Archive, http://classics.mit.edu/Plato/.

8. Gilbert Simondon, *Two Lessons on Animal and Man*, trans. Drew S. Burk (Minneapolis, MN: Univocal, 2012), 43.

9. Ibid., 42.

10. *De anima*, in *The Complete Works of Aristotle: The Revised Oxford Translation*, ed. Jonathan Barnes, vol. 1 (Princeton, NJ: Princeton University Press, 1984), 424a30–35. Hereafter cited in the text as *DA*.

11. Aristotle, *Parts of Animals*, trans. A. L. Peck and E. S. Forster, Loeb Classical Library 323 (Cambridge, MA: Harvard University Press, 1937), 656a1–10. Hereafter cited in the text as *PA*.

12. See Thacker's *After Life*, where he argues that "For Aristotle, *psukhe* is the *arche* of *zoe* and *bios*" (13).

13. On Heidegger's profound debts to Aristotle see Thomas Sheehan, *Making Sense of Heidegger* (New York: Rowman and Littlefield, 2015), 31–110.

14. For excellent discussions of Derrida's obsessive and long-term interest in Heidegger's 1929–30 lecture course see Michael Naas's *The End of the World and Other Teachable Moments* (New York: Fordham University Press, 2014); and David Farrell Krell's *Derrida and Our Animal Others* (Bloomington: Indiana University Press, 2013).

15. Martin Heidegger, *The Fundamental Concepts of Metaphysics: World, Finitude, Solitude*, trans. William McNeill and Nicholas Walker (Bloomington: Indiana University Press, 2001), 179. Hereafter cited in the text as *WFS*.

16. On this point see Simondon's *Two Lessons*, where he argues that for the Greeks, plant, animal, and human life were not so much distinguished by subtending essential differences but relative powers of deployment or "quantities of intelligence, quantities of reason (of nous), the nous of a plant being less strong, less detailed, and less powerful than that of an animal, the nous of the animal itself being less strong, less detailed, and less powerful than that of man" (36).

17. Heidegger here is likely thinking of Aristotle, on sleep and plants: "Nor likewise would anyone desire life for the pleasure of sleep either; for what is the difference between slumbering without being awakened from the first day till the last of a thousand or any number of years, and living a vegetable existence? Any way plants seem to participate in life of that kind; and so do children too, inasmuch as at their first procreation in the mother, although alive, they stay asleep all the time. So that it is clear from considerations of this sort that the precise nature of well-being and of the good in life escapes our investigation" (*Eudemian Ethics* 1.1216a), www.perseus.tufts.edu/hopper/text?doc= Perseus:text:1999.01.0050:book=1:section=1216a.

18. I note in passing that when Heidegger does positively refer to plants and thinking in other places, he tends to do so circuitously, by offering a gloss on someone else's invocation of plant life—as he does for example when invoking "the truth of what Johann Peter Hebel says" at the end of his "Memorial Address" in the *Discourse on Thinking*, trans. John M. Anderson and E. Hans Freund (New York: Harper and Row, 1966): specifically Hebel says, "We are plants which—whether we like to admit it to ourselves or not—must with our roots rise out of the earth in order to bloom in the ether and to bear fruit" (57). Likewise, Heidegger cites and interprets Silesius's "The rose is without why" in *The Principle of Reason*, trans. Reginald Lilly (Bloomington: Indiana University Press, 1996), 35–43). Elsewhere, he quotes and glosses Descartes's letter to Picot, wherein Descartes writes: "Thus the whole of philosophy is like a tree: the roots are metaphysics, the trunk is physics, and the branches that issue from the trunk are the other sciences." ("The Way Back into the Ground of Metaphysics," trans. Walter Kaufmann in his *Existentialism from Dostoevsky to Sartre* [New York: Penguin, 1975], 265). For an attempt to recast Heidegger's philosophy of time in vegetable terms, see Marder's *Plant-Thinking* (96–101).

19. Graham Harman, *Tool Being: Heidegger and the Metaphysics of Objects* (Chicago: Open Court, 2002).

20. See Sheehan's article, "Heidegger's Topics: Excess, Access, Recess." *Tijdschrift voor filosofie* 41, no. 4 (1979): 615–35.

21. On this point consult Krell's tour de force *Derrida and Our Animal Others*, where he argues that Heidegger is not as hard on animals as Derrida makes him out to be in *The Beast and the Sovereign*. Krell goes back through *Being and Time* on this question of "benumbment," finding Dasein's link to animality there, rather than seeing a hard-and-fast distinction: Krell quotes Heidegger from *Sein und Zeit* (113): "At first and for the most part, Dasein is benumbed by its world," a quotation that Krell glosses

like this: "'Dazzled,' 'dazed,' 'benumbed' are all trying to translate *benommen*. Apparently, for much of the time Dasein comports itself in the world precisely in the way animals behave in theirs. Nevertheless, in the 1929–30 lecture course, *Benommensein* will be used to earmark and brand animals as such, animals specifically and exclusively, animals as excluded from all *proper* world-relation, animals as life-and-life-only, just plain life, *Nur-noch-leben*. Presumably, even the dullest of Daseins could never regress to this level. And yet" (109).

Even if animals and Dasein share benumbment, a case Krell convincingly makes— he argues that "such bedazzlement enlarges my life. . . . It is sometimes anxiety, sometimes joy, sometimes profound melancholy that enables us to contemplate our closeness to other forms of life, in spite of all the differences" (159)—there nevertheless remains this tricky problem of parsing out "authentic" (proper) versus "inauthentic" (improper) benumbment, among humans and animals. As Heidegger insists in *Being and Time*, trans. Joan Stambaugh and Dennis Schmidt (Albany, NY: SUNY Press, 2010): "But Angst can arise authentically only in a resolute Dasein. He who is resolute knows no fear, but understands the possibility of Angst as the mood that does not hinder or confuse him. Angst frees him from the 'null' possibilities and lets him become free for authentic ones" (*BT* 316; *SZ* 344).

Now of course everyone who's taken the *Being and Time* seminar knows that "resoluteness" is an accurate, but nevertheless contentious, translation of *Entschlossenheit*— precisely because *resoluteness* suggests the opposite of Heidegger's German, which is probably better rendered as literally "disclosedness" (which is the usual translation of *Erschlossenheit*). In English *resolute* suggests that I'm sure, I know, I'm confident, whereas a literal rendering of *Ent-schlossen-heit* ("dis-closed-ness") far rather suggests "openness," being un-closed, unsure, not-knowing. As with the late Heidegger's gloss of *Ereignis* as *Enteignis* (the event of appropriation is the event of dispersal), the *Ent-* words in Heidegger are often simultaneously *Er-* words: appropriation is dispersion and vice versa. This is the sense that Krell leans heavily upon, all this *Gelassenheit* as letting-be in Heidegger: at his best, Heidegger suggests that all living things share in being bedazzling and bedazzled. However, that positive reading can quickly turn dark when the authentic life of openness or letting be is rethematized as the (only?) proper life—the distinction between those chosen few in tune with being and thinking, and all the inauthentic masses enthralled by technology's presence-to-hand, who can very quickly become rethematized as improper forms of life. Specifically on retranslating *Entschlossenheit*, see Nikolas Kompridis's *Critique and Disclosure: Critical Theory Between Past and Future* (Cambridge, MA: MIT Press, 2006), 58–59. For their part, McNeill and Walker split the difference in *WFS*, and translate *Entscholssenheit* as "resolute disclosedness."

22. As David Krell has pointed out to me, here we could look back to *Being and Time*, section 80, where Heidegger discusses Dasein's everyday temporality also in terms of the sun, opening up a potential commonality among the worlds of Dasein, plants, and animals—everything under the sun: "Everyday *circumspect* being-in-the-world needs the *possibility of sight*, that is, brightness, if it is to take care of things at hand within what is present. With the factical disclosedness of the world, nature has been discovered for

Dasein. In its thrownness Dasein is subject to the changes of day and night. Day with its brightness gives the possibility of sight. . . . When the sun rises, it is time for . . . [second ellipsis Heidegger's]" (BT 392–93).

23. On this central issue of legitimate and illegitimate life-forms in Heidegger see Campbell's incisive summary in *Improper Life*: "Heidegger has essentially created a new grouping where before only mankind or humanity existed, configuring two forms of life where there was only one who acted (and wrote) properly; the first is given over entirely to the social, technical world of metaphysical passion, whereas the other . . . is he who must be saved. . . . A divide comes to separate Western historical man, who sits astride the precipices of history—and therefore who must be saved, given the immediate danger—and those who, in attempting to master technology, become its subject" (9–10).

24. *Gesamtausgabe* 94, 408, translated and cited in Gregory Fried, "The King Is Dead: Heidegger's *Black Notebooks*," *Los Angeles Review of Books*, 13 Sept. 2014, http://lareviewofbooks.org/review/king-dead-heideggers-black-notebooks.

25. Ibid.

26. Ibid.

27. David Farrell Krell, "Heidegger's *Black Notebooks*, 1931–41." *Research in Phenomenology* 45, no. 1 (2015): 127–60. Krell translates the specific passage: "One of the most hidden figures of the gigantic and perhaps the oldest one is the tenacious skillfulness in reckoning and manipulating and meddling in which the worldlessness of Jewish civilization is grounded" (134).

28. Roberto Esposito, *Communitas*, trans. Timothy Campbell (Stanford, CA: Stanford University Press, 2009), 99.

Chapter 3. Animal and Plant, Life and World in Derrida

1. Jacques Derrida, *The Animal That Therefore I Am*, trans. David Wills (New York: Fordham University Press, 2008). Hereafter cited in the text as ATT.

2. See "Animal Studies Cross Campus to Lecture Hall," 2 Jan. 2012, www.nytimes .com/2012/01/03/science/animal-studies-move-from-the-lab-to-the-lecture-hall.html.

3. Jacques Derrida, "White Mythology: Metaphor in the Text of Philosophy," in *Margins of Philosophy*, trans. Alan Bass (Chicago: University of Chicago Press, 1984), 250. Hereafter cited in the text as "WM."

4. Derrida here cites Aristotle's *Metaphysics* (1006a12–15) on the irreducibility of noncontradiction: "it is absurd to reason with one who will not reason about anything, insofar as he refuses to reason. For such a man, as such, is seen already to be no better than a mere vegetable."

5. While I don't want to deny that "White Mythology" takes up the question of animality—it most certainly does—I'm just marking here that the essay is also seriously engaged with vegetal life and the status of metaphor. The essay begins, "From philosophy, rhetoric. That is, here, to make from a volume, approximately, more or less, a flower, to extract a flower, mount it, or rather to have it mount itself, bring itself to light—and turning away, as if from itself, come round again, such a flower engraves" ("WM" 209).

6. Jacques Derrida, *Glas*, trans. John P. Leavey (Lincoln: University of Nebraska Press, 1985), 31b. Hereafter cited in the text as *G*.

7. Later, Heidegger goes on to state quite straightforwardly that "what we call here world-formation is also the ground of the very inner possibility of the logos" (*WFS* 335).

8. Jacques Derrida, *Chaque fois unique, la fin du monde*, ed. Pascale-Anne Brault and Michael Naas (Paris: Éditions Galilée, 2003), 9.

9. Martin Hägglund, *Radical Atheism: Derrida and the Time of Life* (Stanford, CA: Stanford University Press, 2008), 96.

10. Jacques Derrida, *The Work of Mourning*, ed. and trans. Pascale-Anne Brault and Michael Naas (Chicago: University of Chicago Press, 2001), 115. See in this same text Derrida's thoughts in his letter to Max Loreau's widow: "I lack the strength to speak publicly and to recall each time another end of the world, the same end, another, and each time is nothing less than an origin of the world, each time the sole world, the unique world, which, in its end, appears to us as it was at the origin—sole and unique—and shows us what it owes to the origin, that is to say, what will have been, beyond every future anterior" (95). See also Derrida's recollection for his friend Jean-Marie Benoist: "death takes from us not only some particular life within the world, some moment that belongs to us, but each time, without limit, someone through whom the world, and first of all our own world, will have opened up in a both finite and infinite—mortally infinite—way" (107).

11. See Jacques Derrida, *Sovereignties in Question: The Poetics of Paul Celan*, trans. Thomas Dutiot (New York: Fordham University Press, 2005), esp. 141–60, as well as numerous places throughout *The Beast and the Sovereign* lectures, for his extensive commentary on this line from Celan.

12. My thanks to Michael Naas, from the Derrida Translation Project, for offering me this citation from the not-yet-published second year of Derrida's death penalty seminar. Naas cites it in his *The End of the World and Other Teachable Moments: Jacques Derrida's Final Seminar* (New York: Fordham University Press, 2014), chap. 2n14.

13. This quotation is from Gasché's endorsement of Hägglund's *Radical Atheism*, on the back cover.

14. It will have to remain for some future project to see how this late Derridean emphasis on life, death, and world squares with Derrida's (as yet unpublished) 1975–76 lecture course on "Life Death," which was in part concerned with revisiting "general textuality" and its relations to biology (mostly understood through the work of Francois Jacob). In his essay "The Text and the Living: Derrida Between Biology and Deconstruction"—*Oxford Literary Review* 36, no. 1 (2014): 95–114—Francesco Vitale cites and translates Derrida from the archive of that course:

> When the first event, the real origin etc., is a text, has the structure of a text, this fabulous adventure can always reproduce itself. This is what happens with the living being if it has the structure of text. . . . Naturally, this textual self-reference, this closure onto itself of the text that refers only to text, has nothing tautological or autistic about it. On the contrary. It is because alterity is irreducible here that there is only text. It is because no term, no element here has any self-sufficiency nor any effect that does not refer to

the other and never to itself that there is text; it is because the set called text cannot close onto itself that there is only text, and that the so-called "general text" (an obviously dangerous and merely polemical expression) is neither a set nor a totality. (109) In short, Derrida here argues that "the living is a text" (111).

In "The Feeling of Life"—*Oxford Literary Review* 36, no. 1 (2014): 49–62—Alexander Garcia Duttman offers a way of connecting the "Life Death" seminar of 1975–76 to Derrida's later refrain of "each time unique, the end of the world." Duttman writes, "If one accepts . . . the argument about iterability as an argument about life, then one could perhaps infer from it that life, for the deconstructionist, is the occurrence, or the event, of what always occurs only once because it occurs more than once, and of what always occurs more than once because it occurs only once. One could infer from it that life cannot be told apart from a feeling of life, from a certain expressive intensity, since life is tension, a tension between a 'once-only' and a 'once-more'?" (57). In other words, the only-once, life's uniqueness, is made possible and iterable only by its entanglement with (and mediation by) the general text or the trace: the event or "world" of life, each time unique, is opened solely through its ruination (the end of the world), which constitutes the subject's "feeling" or experience of life-death.

15. Martin Hägglund, "The Challenge of Radical Atheism: A Response," *New Centennial Review* 9, no. 1 (2009): 245–46.

16. The Derridean world in this sense is indeed a "fiction" or a "phantasm." As Derrida writes, "it seems to be as if we were behaving as if we were inhabiting the same world and speaking the same language, when in fact we well know—at the point where the phantasm precisely comes up against its limit—that this is not true at all" (*BS* 2:268). The world, in other words, is a kind of enabling fiction or projection, one that keeps at bay what Derrida calls the "infantile but infinite anxiety that *there is not the world*" (*BS* 2:83). As Michael Naas explains in *Derrida from Now On* (New York: Fordham University Press, 2009), "The [Derridean] phantasm can thus be described as a projection on the part of the subject that is then taken to be something external to the subject, a projection that then has real effects . . . and affects that then reinforce the phantasm" (255). As such, then, the phantasm is at the origin of the notion of world, as a kind of necessary fiction that, because of its necessity to the social world, can never be debunked. As Naas points out, for Derrida "the phantasm is not an error to be measured in relation to truth; it is not some imitation, image, or representation to be measured against the real but is akin to what Freud, in *The Future of an Illusion* and elsewhere, terms an 'illusion.' Not a representation or misrepresentation of the way things are but a projection on the part of a subject or nation-state of the way one would wish them to be—and thus, in some sense, the way they become, with all their real, attendant effects" (207).

17. This sense of life as living-on or survival remains crucial to Derrida right up to the end. See his *Learning to Live Finally: The Last Interview*, trans. Pascale-Anne Brault and Michael Naas (Hoboken, NJ: Melville House, 2007): "I have always been interested in this theme of survival, the meaning of which is *not to be added on* to the living and dying. It is originary: life *is* living on, life *is* survival [la vie *est* survie]. To survive in the usual sense of the term means to continue to live, but also to live *after* death. . . . All the

concepts that have helped me in my work, and notably that of the trace or of the spectral, were related to this 'surviving' as a structural and rigorously originary dimension" (26).

18. Sean Gaston, *The Concept of World from Kant to Derrida* (Lanham, MD: Rowman and Littlefield, 2013), 133.

19. Timothy Clark, "What on World Is the Earth?" *Oxford Literary Review* 35, no. 1 (2013): 5–24, 18.

20. On childbirth as absolute arrival see Derrida's "Artifactualities," in *Echographies of Television* (London: Wiley, 2002), 20; on originary technicity see Arthur Bradley's *Originary Technicity: The Theory of Technology from Marx to Derrida* (New York: Palgrave, 2011).

21. See the NIH Human Microbe Project, www.nih.gov/news/health/jun2012/nhgri-13.htm. For some thoughts on how attending to microbial life might change philosophical discourse, see Maureen A. O'Malley's *Philosophy of Microbiology* (Cambridge: Cambridge University Press, 2014): "Overall, microbiology makes philosophers and biologists confront important ontological issues about the adequacy of a focus on single organisms and lineages, and encourages them to explore whether collaborative adaptive and otherwise coevolving units might in fact be the appropriate focus of . . . study" (13). See also the essays collected in Steven Shaviro's *Cognition and Decision in Non-human Biological Organisms* (Ann Arbor, MI: Open Humanities Press, 2011).

22. Naas, *Derrida from Now On*, 208.

23. The penultimate paragraph in the Hegel column is the "final" one insofar as the very last section is in fact the beginning. Page 1 is a continuation from the final page, and the column begins in mid-sentence, "what, after all, of the remain(s), today, for us, here, now, of a Hegel?" (1a). The penultimate (though secretly final) paragraph of the Hegel column concludes: "The syllogism of spiritual art (epic, tragedy, comedy) leads esthetic religion to revealed religion. Through comedy then" (*G* 262a).

24. Claudette Sartiliot, *Herbarium/Verbarium: The Discourse of Flowers* (Lincoln: University of Nebraska Press, 1993), 152. I thank Peggy Kamuf for calling my attention to this extraordinary book.

25. When asked by Derek Attridge what he means by literature, or what it meant to him to be invested in literature as a young man, Derrida answers: "literature seemed to me, in a confused way, to be the institution which allows one to say everything, in every way. . . . The institution of literature in the West, in its relatively modern form, is linked to an authorization to say everything, and doubtless too to the coming about of the modern idea of democracy" (in Derrida's *Acts of Literature*, ed. Derek Attridge [London: Routledge, 1991], 37).

26. Ibid., 73.

27. Hegel, *Phenomenology of Spirit*, trans. A. V. Miller (Oxford: Oxford University Press, 1976), 4.

28. Ibid., 9. For more on vegetable life in Hegel, and more generally in German romanticism, see Miller's *Vegetative Soul*; and chap. 11 of David Farrell Krell's *Contagion* (Bloomington: Indiana University Press, 1998).

29. Krell expertly glosses the word in *Derrida and Our Animal Others*: "as a verb of being, *Walten* resists conjugation: it appears only in the third person singular or plural

indicative; it often seems to perform as an impersonal verb, taking a neuter pronoun, therefore reminiscent of the *es gibt*, the 'granting' of time and being. Neither you nor I can *walten*, nor even he or she, certainly not Heidegger, nor even the sovereign himself or herself can *walten*, unless he or she be God" (110).

30. Quentin Meillassoux, *After Finitude*, trans. Ray Brassier (London: Bloomsbury, 2010), 5.

31. Slavoj Žižek, *Less Than Nothing* (London: Verso, 2012), 642. It's ironic that Graham Harman, who's pretty much the brand manager of Object-Oriented Ontology and Speculative Realism, explicitly names not Derrida but Žižek as his primary example of correlationist thinking: "Most recent philosophy in the continental tradition can safely be described as a philosophy of access to the world. Concurring with the spirit of Žižek's principle that 'Kant was the first philosopher,' it assumes that the human-world gap is the privileged site for all rigorous philosophy" (Graham Harman, *The Quadruple Object* [Winchester, England: Zero Books, 2011], 136). Harman ends up issuing "a call for an escape from the closed circle of Žižek and his confederates" (62).

32. Ian Bogost, *Alien Phenomenology; or, What It's Like to Be a Thing* (Minneapolis: University of Minnesota Press, 2012), 31.

33. Levi Bryant, Nick Srnicek, and Graham Harman, introduction to *The Speculative Turn: Continental Materialism and Realism* (Victoria, Australia: re-press, 2011), 3.

34. Graham Harman, "On the Undermining of Objects," in Bryant, Srnicek, and Harman, *The Speculative Turn*, 21–40, 25.

35. For an attempt to reconcile Derrida's thought with "speculative realism" see Michael Marder's *The Event of the Thing: Derrida's Post-Deconstructive Realism* (Toronto: University of Toronto Press, 2011).

36. Jacob von Uexküll, *Theoretical Biology* (Boston: Harcourt Brace, 1926), xv.

37. Thacker, *After Life*, x.

38. Henry Staten, "Derrida, Dennett, and the Ethico-political Project of Naturalism," *Derrida Today* 1, no. 1 (2008): 19–41, 39.

Chapter 4. From the World to the Territory

1. Miller, *Vegetative Soul*, 185.

2. My posing of this question is nested within and schooled by Gregg Lambert's posing of the same question in his *In Search of a New Image of Thought*, where he shows (quite persuasively) how rhizomatics grows out of Deleuze's reading of Proust combined with Guattari's thinking about the "neuro-vegetative" diagram of the brain. He argues that rhizomatics goes beyond the question of territory, toward nothing less than a new image for thought; see especially 43–61.

3. Gilles Deleuze and Felix Guattari, *A Thousand Plateaus: Capitalism and Schizophrenia*, trans. Brian Massumi (Minneapolis: University of Minnesota Press, 1987), 2:22. Hereafter cited in the text as *TP*.

4. Gilles Deleuze, *Difference and Repetition*, trans. Paul Patton (New York: Columbia University Press, 1995), xvii. Hereafter cited in the text as *DR*.

5. For happy exceptions to this general rule that rhizomatics is about anything

but plants, see especially Lambert's *In Search of a New Image of Thought*; and Karen Houle's fine "Animal, Vegetable, Mineral: Ethics as Extension or Becoming? The Case of Becoming-Plant," *Journal for Critical Animal Studies* 9, no. 1–2 (2011): 89–116; and Laura Marks's "Vegetable Locomotion," in *Revisiting Normativity with Deleuze*, ed. Rosi Braidotti and Patricia Pisters (London: Continuum, 2012), 203–17); as well as, of course, Marder's *Plant-Thinking*.

6. Gilbert Simondon, "The Position of the Problem of Ontogenesis," trans. Gregory Flanders, *Parrhesia* 7 (2009): 4–16, 6 (Simondon's emphasis).

7. Ibid., 7.

8. Ibid., 9.

9. John Protevi, *Life, War, Earth: Deleuze and the Sciences* (Minneapolis: University of Minnesota Press, 2013), 149.

10. In *Vegetable Soul* Miller critiques rhizomatics merely as a metaphor: "it seems necessary to question any pairing that relies—as Deleuze and Guattari's account ultimately does—on the traditional definition of metaphor as a lively description that illuminates the structure of another reality that lies behind it, in this case, the description of plants that stands in for the description of writing" (186).

11. Marks, "Vegetable Locomotion," 214.

12. Marder, *Plant-Thinking*, 84.

13. See Shaviro's blog post of 18 Dec. 2004, where he argues that, given recent discoveries concerning plant intelligence, "Deleuze and Guattari's distinction between 'rhizomatic' and 'arborescent' modes of organization needs to be rethought. In point of fact, trees are far less binaristic and hierarchical than Deleuze and Guattari make them out to be" (www.shaviro.com/Blog/?p=376). See also Marder's *Plant-Thinking* for a different version of this critique.

14. David Wood, "Truth and Trees; or, Why We Are All Really Druids," in *Rethinking Nature: Essays in Environmental Philosophy*, ed. Bruce V. Foltz and Robert Frodeman (Albany, NY: SUNY Press, 2004), 38.

15. Kari Weil, *Thinking Animals: Why Animal Studies Now?* (New York: Columbia University Press, 2012), 24, 149.

16. Marder, *Plant-Thinking*, 85.

17. Lambert, *In Search of a New Image of Thought*, 52.

18. Ansell-Pearson, *Germinal Life*, 95.

19. Marks, "Vegetable Locomotion," 206.

20. Ansell-Pearson, *Germinal Life*, 96.

21. Deleuze, *Negotiations*, 143.

22. See Claire Colebrook, *Deleuze and the Meaning of Life* (London: Continuum, 2011); and Protevi's *Life, War, Earth*.

23. From "A Is for Animal," in the taped interview *Gilles Deleuze from A to Z*, with English subtitle translation by Charles Stivale (Cambridge, MA: MIT Press, 2011), DVD.

24. Protevi, *Life, War, Earth*, 164.

25. Elaine Miller, "Vegetable Genius: Plant Metamorphosis as a Figure for Thinking and Relating to the Natural World in Post-Kantian German Thought," in *Rethinking*

Nature: Essays in Environmental Philosophy, ed. Bruce V. Foltz and Robert Frodeman (Albany, NY: SUNY Press, 2004), 116.

26. Deleuze and Guattari, *What Is Philosophy?* trans. Hugh Tomlinson and Graham Burchell (New York: Columbia University Press, 1994), 212. Hereafter cited in the text as *WP*.

27. For all of Žižek's insisting on this point—the claim is made as early as *Tarrying with the Negative* and is all over *Organs Without Bodies*—at this Google-Earth level of argumentation (D&G's ontological stance of flows and discontinuities works to naturalize the everyday practices of capitalism), it's not clear how Žižek's primary commitments to the ontology of Lacanian lack and desire don't fall prey to the same argument. If for Lacan we are never satisfied with what we have, if desire always wants more and is constantly fooled, and in addition there's no way out of this cycle (it is to some degree "life" itself), this seems to provide a very convincing ontological or naturalizing explanation concerning why we all want and need to buy stuff and always require new stuff: if our unconscious is structured ontologically by endless desire to fill a lack, isn't capitalism the logical social system to complement that "real"?

28. Deleuze, "Postscript on the Societies of Control," *October* 59 (Winter 1992): 3–7, 3.

29. Ibid., 4.

30. Foucault, *Discipline and Punish*, 211.

31. Michel Foucault, *Dits et écrits*, ed. Daniel Defert and Francois Ewald (Paris: Gallimard, 1994), 4:662.

32. This I think constitutes the Foucauldian-Deleuzean answer to Thomas Frank's question in *What's the Matter with Kansas?* (New York: Holt, 2002). There, Frank wonders why poorer rural people in the United States vote for conservatives and primarily focus their attention on social issues (abortion, gun control, immigration, gay marriage, government health care) rather than voting for their own economic self-interest, which presumably would be better served by left-leaning candidates. Biopower as control suggests a relatively simple answer to that question: life-issues are at present the biopolitical territory within which all other issues, including economic ones, are adjudicated. Politically speaking, biopower shows how questions of life and lifestyle coemerge anew with the terrain of politics as well as economics. Again, it's less the world that's changed than the territory.

33. Foucault, *The Birth of Biopolitics*, 259–60.

34. Think of Deleuze's work on stupidity in *Nietzsche and Philosophy*, trans. Hugh Tomlinson (New York: Columbia University Press, 1982): stupidity is not the binary opposite of knowledge, but knowledge is simply that which injures stupidity, keeps it from taking over the entire territory of thought and action: "Great as they are, stupidity and baseness would be greater still if there did not remain some philosophy which always prevents them from going as far as they would wish" (106).

35. Simondon, *Two Lessons*, 32.

Coda

1. As Dawne McCance notes in *Critical Animal Studies: An Introduction* (Albany, NY: SUNY Press, 2013), despite their substantial differences, "both Regan and Singer re-

sort to an anthropomorphic 'like us' standard in determining which nonhuman animals count as having equal interests or rights" (39).

2. Quoted in Richard Knight, "Biodiversity Loss: How Accurate Are the Numbers?" *BBC News Magazine*, 24 April 2012, www.bbc.com/news/magazine-17826898.

3. Cited in Kathryn Lougheed, "There Are Fewer Microbes Out There Than You Think," *Nature*, 27 August 2012, www.nature.com/news/there-are-fewer-microbes-out -there-than-you-think-1.11275. As O'Malley points out in *Philosophy of Microbiology*, almost everything that presently lives—or has ever lived—on this planet is a bacterium (see 208–12).

4. These exact numbers are of course hard to come by and hotly contested. See the Center for Biodiversity's website (which puts the number at "dozens" each day): www.biologicaldiversity.org/programs/biodiversity/elements_of_biodiversity/ex-tinction_crisis/. Less controversial is the role of humans in accelerating this process of other species' extinction: best estimates put Anthropocene extinction rates at one thousand times the natural "background rate"—which is to say, the rate of extinction that would have happened regardless of human domination of the global scene. See Christine Dell'Amore, "Species Extinction Happening 1,000 Times Faster Because of Humans?" *National Geographic*, 29 May 2014, http://news.nationalgeographic.com/news/2014/05/140529-conservation-science-animals-species-endangered-extinction/.

5. Wolfe, *Before the Law*, 19.

6. Ibid., 103.

7. Jacques Derrida, "Structure, Sign, and Play in the Discourse of the Human Sciences," in *Writing and Difference*, trans. Alan Bass (Chicago: University of Chicago Press, 1978), 292.

8. As Derrida puts it in "The Deconstruction of Actuality": "There would be no event, no history, unless a 'come hither' opened out and addressed itself to someone, to someone else whom I cannot and must not define in advance—not as subject, self, consciousness, nor even as animal, God, person, man or woman, living or dead" ("The Deconstruction of Actuality: An Interview with Jacques Derrida," trans. Jonathan Rée, *Radical Philosophy* 68 [Autumn 1994]: 28–41, 32).

9. See Lauren Morello, "Phytoplankton Population Drops 40 Percent Since 1950," *Scientific American*, 29 July 2010, www.scientificamerican.com/article/phytoplankton -population/.

10. Wolfe, *Before the Law*, 51.

11. Marie-Monique Robin, *The World According to Monsanto: Pollution, Corruption, and the Control of the World's Food Supply*, trans. George Holoch (New York: New Press, 2010), 203.

12. See www.ers.usda.gov/data-products/adoption-of-genetically-engineered-crops-in-the-us/recent-trends-in-ge-adoption.aspx.

13. Cited in Robin, *World According to Monsanto*, 205–6).

14. Cited in Dreyfus and Rabinow, *Michel Foucault*, 187.

15. Morton, *Ecology Without Nature*, 20.

16. For an overview of these "what about anthrax?" objections to Esposito's robust sense of life, as well as an incisive response, see Mitchell's *Experimental Life*, 220–26.

17. Wolfe, *Before the Law*, 59.

18. Ibid., 93.

19. This is Deleuze from *The Logic of Sense* (103), quoted approvingly by Esposito in *Bios*, 193.

20. Esposito, *Bios*, 186.

21. Ibid., 187.

22. Houle, "Animal, Vegetable, Mineral," 111–12. I also thank Houle for drawing my attention to the quotation from TP that follows.

23. Wolfe, *Before the Law*, 56.

24. For some thoughts on Derrida and climate change see the essays collected in Tom Cohen's *Telemorphosis: Theory in the Era of Climate Change* (Ann Arbor, MI: Open Humanities Press, 2012). For the sharpest contemporary thinking about extinction see Colebrook's *Death of the Post-Human*.

25. On the latest in genetic manipulation and life (which is to say, the ability to manipulate genes to create life rather than merely to "read" the life-information contained in DNA), see Ross Thyer and Jared Ellefson, "Synthetic Biology: New Letters for Life's Alphabet," *Nature*, 15 May 2014, 291–92, www.nature.com/nature/journal/vaop/ncurrent /full/nature13335.html; and Denis A. Malyshev et. al, "A Semi-synthetic Organism with an Expanded Genetic Alphabet," *Nature*, 15 May 2014, 385–88, www.nature.com/nature/ journal/vaop/ncurrent/full/nature13314.html.

26. See Derrida's essays on apocalypse, "Of an Apocalyptic Tone Recently Adopted in Philosophy," *Oxford Literary Review* 6, no. 2 (1984): 3–37; and "No Apocalypse, Not Now," *diacritics* 14, no. 2 (1984): 20–31.

27. See Mark Fisher, *Capitalist Realism: Is There No Alternative?* Winchester, England: Zero Books, 2009.

28. See Jameson's "Future City," *New Left Review* 21 (May–June 2003): 65–79, 76.

29. Michel Foucault, *Speech Begins After Death*, trans. Robert Bononno (Minneapolis: University of Minnesota Press, 2013), 45.

30. Thacker, *After Life*, 5. Thacker continues: "When life is taken as subjective experience, life is projected from subject to object, self to world, and human to non-human. Another name for this process is anthropomorphism," with its concomitant notion of "life that is rooted in a living, experiencing subject. And, since a reflexive awareness of living is implied in the very idea of life as experience, this also means that life becomes a human-centric idea" (5).

INDEX

Adorno, Theodor, 3, 17, 125n
Agamben, Giorgio, xi, 11, 14–27, 51, 84,
    116, 119, 128n–129n
Animal Studies, ix–xv, 1–28, 49–61,
    109–122
Ansell-Pearson, Keith, 94–95, 124n, 137n
Anthropocene, 74, 113, 139n
Aquinas, St. Thomas, xii, 37, 123n
Arendt, Hannah, 16
Aristotle, 9, 11–12, 25, 29–37, 50, 60, 68, 95,
    98, 126n, 129n–130n, 132n

Beckett, Samuel, 90
Biopower, ix–x, xv, 1–27, 31, 36–37, 60, 84,
    93, 96, 101–107, 113–122, 138n
Bogost, Ian, 76, 136n
Brown, Wendy, xii, 26
Bryant, Levi, 76, 136n
Burroughs, William, 90, 102
Butler, Judith, xii, xv, 110, 114, 124n

Cain and Abel, 50–51
Campbell, Timothy, 118, 124n, 132n
Capitalism, 16–24, 100–107, 112–114,
    121–122, 127n, 138n
Carstensen, Peter, 113
Celan, Paul, 55, 73
Chamovitz, Daniel, xiii–xiv, 124n, 130n
Clark, Timothy, 58, 135n
Cohen, Jeffrey Jerome, xiii, 123n
Cohen, Tom, 140n

Colebrook, Claire, xiii, xvii, 96, 123n,
    124n, 137n, 140n
Correlationism, 75–81

Darwin, Charles, 8
Dean, Tim, 24, 128n
Deleuze, Gilles, 19, 22, 84, 100–107, 127n,
    128n
Deleuze, Gilles and Felix Guattari, x–xi,
    xv, 31, 56, 83–100, 114, 118–119, 121,
    124n, 136n–138n
Derrida, Jacques, x–xi, xv, 1, 11–14, 15, 29,
    31, 37, 43, 45, 49–81, 83–84, 96, 110–111,
    116–117, 119–120, 121, 124n, 127n, 128n,
    129n, 132n–136n, 139n, 140n
Doyle, Richard, v, xiii, xviii, 56, 123n,
    124n, 126n
During, Simon, 125n
Duttman, Alexander Garcia, 134n

Eliot, T.S., 90
Esposito, Roberto, 47, 116–118, 124n, 132n,
    140n
Extinction, xv, 106–107, 109–111, 120–122,
    141n, 142n

Fisher, Mark, 121, 140n
Foer, Jonathan Safran, 49
Foucault, Michel, ix–x, xv, 1–27, 31, 33, 64,
    80, 84, 91, 93, 98, 101–107, 113–114, 116,
    118–122. 124n–128n, 138n, 140n

Francione, Gary, xi–xii
Frank, Thomas, 138n
Freud, Sigmund, 8, 95

Gasché, Rodolphe, 56
Gaston, Sean, 57–58, 135n
Gates, Henry Louis, 114
Genet, Jean, 60–71
Genetically Modified Organisms
    (GMOs), xiii, 23, 112–113, 139n

Hagglund, Martin, 54–58, 124n, 133n,
    134n
Hall, Matthew, xiii, 123n
Halle, Francis, xiii, 123n
Haraway, Donna, 3, 11, 125n
Harman, Graham, 41, 76, 130n, 136n
Hauter, Wenonah, xiii, 123n
Hegel, G.W.F, 8, 60–74, 75, 88, 120, 121,
    135n
Heidegger, Martin, xi, 11, 37–47, 51–56,
    71–74, 77–78, 84, 98, 124n, 129n–132n,
    133n, 136n
Hemingway, Ernest, 88
Houle, Karen, 119, 137n, 140n
Hubbel, Stephen, 109

Jameson, Fredric, 114, 121, 140n
Joyce, James, 90

Kamuf, Peggy, 135n
Kant, Immanuel, 75–79, 101, 136n
Kaufmann, Frederick, xiii, 123n
Kelley, Theresa, xiii, 123n
Kohn, Eduardo, xiii, 123n
Krell, David Farrell, xvii, 46, 124n, 129n,
    130n–132n, 135n–136n

Lambert, Gregg, xvii, 92, 136n, 137n
Lawlor, Leonard, 126n–127n
Levinas, Emmanuel, 41

Mallarmé, Stéphane, 90

Marder, Michael, xi, xiii, 12, 31, 87–88, 91,
    123n, 126n, 129n, 130n, 136n, 137n
Marks, Laura, 87, 95, 137n
Marx, Karl, 16, 114
McCance, Dawn, 138n–139n
Meillassoux, Quentin, 75–81, 136n
Microbial life, 57–59, 109, 113, 116–119, 120,
    135n, 139n
Miller, Elaine, xiii, 83, 98, 123n, 135n, 136n,
    137n
Mitchell, Robert, xiii, 124n, 140n
Monsanto Corporation, 112–113
Morton, Timothy, xiii, 114–115, 118, 123n,
    139n

Naas, Michael, xvii, 61, 71–72, 129n, 133n,
    134n, 135n
Nazism (and Heidegger), 45–47
Negri, Antonio, 26, 128n
Nietzsche, Friedrich, 8, 90

O'Hara, Frank, 10, 125n
O'Malley, Maureen, 135n, 139n

Physis (nature), 52, 59–60, 67–74, 78–81
Plant Intelligence, xii–xv, 12, 30–31, 111,
    123n, 126n, 129n, 130n, 137n
Plato, 12, 29–32, 45, 95, 128n, 129n
Pollan, Michael, xiii, 123n, 124n
Protevi, John, xvii, 87, 96–97, 124n, 136n
Psukhe (soul), 31–39, 46, 68, 98,106, 129n

Rabinow, Paul and Hubert Dreyfus,
    127n–128n, 139n
Rhizome, 83–101
Robin, Marie-Monique, xiii, 123n
Rodowick, D.N., 114
Rose, Nikolas, 124n

Sacks, Oliver, xiii, 123n
Sartiliot, Claudette, 62, 135n
Schumpeter, Joseph, 8
Shakespeare, William, 37, 52

Shaviro, Steven, 88, 135n, 137n
Sheehan, Thomas, 42, 129n, 130n
Shukin, Nicole, 4, 125n
Simondon, Gilbert, 31–32, 85–87, 107, 118, 129n, 130n, 137n, 138n
Speculative Realism, 75–81, 136n
Srnicek, Nick, 76, 136n
Staten, Henry, 80, 136n

Taylor, Chloe, 125n
Thacker, Eugene, xv, 78, 122, 124n, 126n, 129n, 136n, 140n
Thierman, Stephen, 4
Trewavas, Anthony, xiii, 123n, 126n, 129n

Uexküll, Jakob von, 42–43, 76–78, 96, 136n

Vegetarianism, xi–xii, 27
Vitale, Francesco, 133n–134n

Weil, Kari, 91, 137n
Wolfe, Cary, xi–xii, xvii, 59–60, 110–112, 118–119, 121, 123n, 125n, 139n, 140n
Wood, David, 88, 137n
World and life, xi, 37–47 (Heidegger), 49–81 (Derrida), 86–100 (Deleuze and Guattari)

Žižek, Slavoj, 75, 101, 136n, 138n

The authorized representative in the EU for product safety and compliance is:
Mare Nostrum Group
B.V Doelen 72
4831 GR Breda
The Netherlands

www.ingramcontent.com/pod-product-compliance
Lightning Source LLC
Chambersburg PA
CBHW020555270326
41927CB00006B/853